The Bad Faith in the Free Market

Peter Bloom

The Bad Faith in the Free Market

The Radical Promise of Existential Freedom

Peter Bloom
Open University
Milton Keynes, UK

ISBN 978-3-319-76501-3   ISBN 978-3-319-76502-0   (eBook)
https://doi.org/10.1007/978-3-319-76502-0

Library of Congress Control Number: 2018936141

Cover Credit : Chaichan Ingkawaranon / Alamy Stock Vector

Printed on acid-free paper

This Palgrave Macmillan imprint is published by the registered company Springer International Publishing AG part of Springer Nature
The registered company address is: Gewerbestrasse 11, 6330 Cham, Switzerland

Preface

A Preliminary Intervention: Confronting Our Bad Faith

It has taken less than two decades since the start of the new millennium for the end of capitalist history to begin. The idea that liberal democracy would reign supreme and free markets would rule the globe is crumbling fast. In its place is a popular revolt that filled with both progressive light and retrogressive shadows. Perhaps the most crucial question of this transitional era is whether we can once more have the courage to reimagine our world in theory and practice. Or will we sacrifice the potential to create a radically new society on the altar of old ideologies or impassioned desires for destruction for its own sake?

At the heart of these urgent and fundamental questions is whether we have the courage to move on from a bad faith in the free market. What though is precisely meant by bad faith? For the famous existentialist French philosopher Sartre—who coined the term—it stands for more than simply believing in a false objective truth. It was the maintaining of this belief despite our knowledge that it was indeed not worthy of such idolatry (as nothing in fact is). It is a deep and often brushed aside form of personal and collective self-deception, the embrace of a divine force to direct our lives even after it has become all too readily apparent that this

God does not exist. It is a bad faith in that it is a continual rejection of the faith in our freedom to choose the existence we desire, a forsaking of our agency to transform our reality.

In the present era, there is the danger of our intensifying our faith in a free market system that clearly does not deserve it. Despite decades of experts publicly extolling its objectivity and inevitability, the 2008 near global meltdown represented a profound existential crisis for capitalism. It supposed inherent meaning, its infallible reflection of human nature, was in an almost an instant torn asunder and revealed to be hollow. The market emperor was shown firmly and finally to be wearing no clothes. And yet our belief in it persists for so many, our embrace of austerity as a cure-all ticket to economic recovery, our faith that with just a few tweaks we could hold at bay our looming economic and social catastrophe.

This book is not a naïve call for us to merely stop believing in capitalism—as if the abstract rejection of the free market would be enough to concretely give birth to a different and better society. By contrast, it is to highlight the importance of recapturing our existential freedom to shape our historical destiny. It asks why we continue to take the free market or any system so "seriously". In the words of Sartre (1956: 627), we must

> repudiate the spirit of seriousness. The spirit of seriousness has two characteristics: it considers values as transcendent givens independent of human subjectivity, and it transfers the quality of "desirable" from the ontological structure of things to their simple material constitution.

Instead of searching desperately for a permanent and universal truth, rather than looking upwards for a God to save us, we should bask in our freedom to create, to experiment, to explore the vast possibilities of our individual and shared existences.

Milton Keynes, UK   Peter Bloom

Contents

1 The Bad Faith in the Free Market: The Need for
Existential Freedom   1

2 Breaking Free from the Free Market: The Existential
Gap of Freedom   19

3 Capitalism's Existential Crisis: Producing
Existential Freedom   41

4 The Facticities of Neoliberalism: Demanding
Existential Freedom   65

5 Capitalist Being and Nothingness: Enjoying
Existential Freedom   91

6 Subjected to the Free Market: The Subject of
Existential Freedom   117

7   Deconstructing the Free Market: The Spectre of
     Existential Freedom   145

8   Reinvesting in Good Faith: The Radical Promise
     of Existential Freedom   171

Index   187

1

The Bad Faith in the Free Market: The Need for Existential Freedom

It seems impossible to even conceive of a non-capitalist society. As the social philosopher Jameson (2003: 76) famously declares, "it is easier to imagine the end of the world than to imagine the end of capitalism." Yet the complete assumption of freedom as exclusively linked to capitalism is being increasingly challenged. The 2008 financial crisis once again brought into sharp relief the limits of market freedom. The dream of meritocracy mixed with personal liberty had turned into a present-day nightmare of rising inequality, economic insecurity, debt bondage, and mass downward mobility. It also raised renewed questions of whether the free market specifically and capitalism generally can provide for a fulfilling personal and social existence.

Emerging from these challenges were fundamental existential concerns. Notably, if the promise of the free market was hollow, then was freedom even possible? Was this truly the "end of history"—a once optimistic claim about capitalism and liberal democracy that had turned into a resigned lament? To this end, the social liberty and personal aspirational impulses previously central to the legitimization of neoliberalism has transformed into an acceptance over its supposed inevitability and deeper almost divine truths about human nature and its possible future. Hence, in place of freedom, free marketers have offered the solace of religion.

© The Author(s) 2018
P. Bloom, *The Bad Faith in the Free Market*,
https://doi.org/10.1007/978-3-319-76502-0_1

1

There was, of course, always an element of the objective and natural about capitalism. It represented human nature at its most pure and essential. It was based on objective economic laws that defied any and all attempts at human control. Yet as the actual rationale for modern capitalism began to falter, its reasonableness weakening in light of actual evidence, the current free market ideology of neoliberalism became increasingly supernatural. It now demanded dogmatic belief from its human followers. Required was an inviolable religious faith in the free market orthodoxy.

A crucial question of our time then is whether we can give up our bad faith in the free market. The degree to which individually and collectively we can dramatically reimagine the meaning and practice of freedom. If we no longer accept that capitalism represents the limit of social possibility, can we wake up from our dogmatic capitalist slumber to embrace and explore new potentialities for our personal and shared existence?

Aim of the Book

This book boldly reconsiders the free market. Innovatively combining existentialist philosophy with cutting-edge post-structuralist and psychoanalytic perspectives, it argues that present-day capitalism has robbed us of our individual and collective ability to imagine and implement alternative and more progressive economic and social systems. To this effect, it has deprived us of our radical freedom to choose how we live and what we can become. In place of this deeper liberty, the free market offers subjects the opportunity to continually reinvest their personal and shared hopes in its dogmatic ideology and policies. This embrace helps to temporary alleviate rising feelings of anxiety and insecurity at the expense of our fundamental human agency. This work exposes our present-day bad faith in the free market and how we can break free from it.

Challenging Freedom

This work attempts to move beyond the existing social limits of market freedom. The goal, in this respect, is to show the concrete limitations and ideological narrowness of currently dominant understandings of liberty and agency associated with the so-called free market. Doing so, though, requires a more complete understanding of what actually is market freedom. What type of liberty does it idealize, how does it seek to practically emancipate us, and what more radical forms of freedom does it perhaps unintentionally gesture towards?

Traditionally, capitalism has defined freedom as... the ability of individuals to, if we are to use the most technical terms, sell their goods and labour freely in an open and competitive market. Using more everyday says language, it is the right to choose one's profession and lifestyle and be fairly rewarded for their labour and enterprise. Quoting the early-nineteenth-century French liberal economist Frederick Bastiat

> Thus, since an individual cannot lawfully use force against the person, liberty, or property of another individual, then the common force—for the same reason—cannot lawfully be used to destroy the person, liberty, or property of individuals or groups.

Echoing these sentiments in the contemporary era, prominent American libertarian politician Ron Paul proclaims:

> It must be remembered that a vast majority of mankind's history has been spent living under the rule of tyrants and authoritarians. The ideas of Liberty are very new when you consider the big picture. By contrast, various forms of socialism and fascism have been adopted over and over again. Be wary of those who try to present these old and tired ideas as something new and exciting. Liberty and free markets are the way forward if we truly desire peace and prosperity.

Significantly, this freedom extends only to opportunity not outcome. It is the right to freely aspire to be successful, according to one's own desires, in a market system. You may try and fail, of course. Yet this failure does not diminish your freedom as it was based on your own lacking qualities not due to any flaw in the market itself. According to influential twentieth-century economist Murray Rothbard, "Free-market capitalism is a network of free and voluntary exchanges in which producers work, produce, and exchange their products for the products of others through prices voluntarily arrived at." Further, no one coerced you into these actions; all that you do or do not achieve is based on your own free will. The contemporary free market economist Jeffery Tucker:

> Even the richest person, provided the riches comes from mutually beneficial exchange, does not need to give anything "back" to the community, because this person took nothing out of the community. Indeed, the reverse is true: Enterprises give to the community. Their owners take huge risks, and front the money for investment, precisely with the goal of serving others. Their riches are signs that they have achieved their aims.

There are, obviously, clear and rather obvious critiques to this narrow version of market-based freedom. Perhaps most notably is that, in practice, this freedom often leads to greater inequality and mass deprivation, conditions that ultimately lessen rather than enhance autonomy. There is also a stronger critique to be raised on purely freedom grounds. It is that wage labour is inherently unfree as it is based on a relationship of economic dependency. This was certainly the view, for instance, of a number of the "founding fathers" of the American Revolution (Foner 1999). This point in and of itself is damaging but not fatal to market freedom as the solution to this dependency problem is economic ownership and entrepreneurism.

A perhaps more damning critique, in this respect, is that this freedom is historical rather than inherent. To this end, it evolved over time to reinforce specific power relations and imbalances. The US political scientist Eric MacGilvray (2011: 1) explores this precise dynamic in his recent book *The Invention of Market Freedom*:

> So complete is this shift in usage that the phrase the "free market" sounds almost redundant to our ears, and the "libertarian" the partisan of liberty is generally understood to be a person who favours the extension of market norms and practices into nearly all areas of life … These dramatic changes in usage are of more than merely historical interest, because freedom over the same period of time become one of the most potent words in our political vocabulary, and the effort to expand the use of the market as a means of realizing social outcomes has greatly intensified, especially in recent decades.

It also highlights a crucial tension for advocates of market freedom, one with definite existential overtures—even if market freedom is the ultimate freedom, can people be free if they are merely conforming to an inscriptive historical discourse? Put differently, can one be socially overdetermined as a free subject and still be considered free as such?

There are also quite serious philosophical challenges to market freedom. In particular, it is unclear whether it is primarily a means or an end? Arguably, its most famous proponent Milton Friedman has argued that it is most definitely the latter. For Friedman and other free-marketers, the government is a necessary evil that must be limited along with efforts to limit—even if well-intentioned—our "economic freedom". He notes "Whether blameworthy or not the use of the cloak of social responsibility, and the nonsense spoken in its name by influential and prestigious businessmen, does clearly harm the foundations of a free society" (Friedman 1970: 1). The creation of a competitive market with minimal regulations or personal limitations, therefore, is the end goal.

In the literature, this view is epitomized by Ayn Rand where she famously, or infamously depending on your perspective, writes of businessman heroes railing against the nefarious attempts of tyrannical governments to restrict the entrepreneurial spirit of free individuals. On the other side of the philosophical divide Sen proposes a capabilities theory in which the market and market mechanisms are just one set of economic freedoms that can serve a population's overall welfare and development. He describes agency in quite empowering and productive terms as "what a person is free to do and achieve in pursuit of whatever goals or values he or she regards as important" (Sen 1985: 2003). For this reason, his approach has been referred to as "people-centred",

> ...which puts human agency (rather than organizations such as markets or governments) at the centre of the stage. The crucial role of social opportunities is to expand the realm of human agency and freedom, both as an end in itself and as a means of further expansion of freedom. The word 'social' in the expression 'social opportunity' ... is a useful reminder not to view individuals and their opportunities in isolated terms. The options that a person has depend greatly on relations with others and on what the state and other institutions do. We shall be particularly concerned with those opportunities that are strongly influenced by social circumstances and public policy... (Drèze and Sen 2002 page 6)

These competing views do point to a deeper philosophical tension plaguing market freedom. It is one that has been widely noted and for good reason. Does one have the right to freely consent or even actively choose to be unfree? Fundamentally, capitalism revolves around selling your labour and indeed your freedom temporarily to the highest bidder. There is thus an ironic dynamic at the heart of the free market—freedom extends only so far as having the right to freely select your servitude. Or more accurately to apply to the servitude that is most attractive to you.

Of course, there is an important additional dimension to this rather ironic form of market freedom. The exchange for temporary unfreedom is not merely the guarantee of material survival. It is also the liberty and resources to pursue happiness outside of work. To this effect, capitalism revolves around what can be referred to as a freedom transaction—the selling of freedom for the concrete possibilities of greater future freedom. Tellingly, this market transaction has recently increasingly spread to employment itself—a job and a career now critical to the achievement of personal fulfilment, not just professional satisfaction.

There is perhaps an immediate temptation to simply dismiss market freedom as a crumbling façade. To throw it unceremoniously onto the ash heap of history given its quite fatal philosophical and empirical failings. Yet, to do so ignores its gesturing towards a deeper existential freedom—the liberty to continually reinterpret and reshape one's existence. The freedom transaction is ultimately at its core an attempt to buy and embody this radical agency, however, temporary and, in the last instance, futile. Consequently, as will be explored the free market was always an existential as much as it was an economic project—one that was similarly hopeful and flawed on both accounts. The next section will examine the snuffing out of the dream of this existential market freedom by the rise of market fundamentalism.

Free Market Evangelists

One of the most striking, and to be fair mocked, figures of the late twentieth and early twenty-first century is the free market evangelical. Almost exclusively an American invention, they preach to their faithful every

Sunday in mega churches and television programmes the social gospel of unrestrained capitalism. At first glance, this is the result of a modern-day Faustian bargain between capitalist and social conservatives. Economic and political elites are more than happy it seems to sell out social liberty for a little soul if it will advance their overall interests.

Of course, there has always been a strong religious dimension to capitalism. Weber (2002) famously wrote about the inexorable relationship of the rise of a market economy with Protestant ethics of frugality and hard work (see also Furnham et al. 1993; Giorgi and Marsh 1990). Yet, in the present age, this capitalist religiosity has taken on seemingly an added fervour. More than simply religion—Christianity in particular—being crucial to the spread of the free market, now capitalism has become a modern religion all onto itself. In the words of Nobel Prize winning economist Joseph Stiglitz (2009: 346), "From a historical point of view, for a quarter of century the prevailing religion of the West has been market fundamentalism. I say it is a religion because it was not based on economic science or historical evidence."

The above discussed growing market fundamentalism reveals how capitalism has entered into a new revivalist phase. Here, the free market is an almost divine force for delivering economic prosperity and political democracy. It alone holds the key to human well-being and individual liberty. Despite mounting evidence to the contrary, the market remains heralded as absolutely critical to our ultimate personal and collective salvation. It reflects an "ineluctability of market forces" based on a "mixture of implicit and hidden assumptions, myths about the history of their own countries' economic development, and special interests camouflaged in their rhetoric of general good" (Kozul-Wright and Rayment 2007: 14).

All that is required is a little faith in the free market's saving grace. Popular shows such as the provocative South Park have already satirically lamented our worship of the "economy" as if it was godlike rather than human controlled. Indeed the more seemingly arbitrary and depersonalized the current financialized economy becomes, the more divinely inspired it appears to be (see Jenkins 2001). All we can do, it seems, is hope that the economy will recover and once again not smite us with the scourge of greater recession.

Personally, this market faith demands a recommitment to making one-self worthy of capitalist salvation. Akin to the evangelical call to "get good with God" through stronger belief and better living, the new market gospel is personal improvement to appease our free market Lord. Hence "neoliberal discourses", according to Konings (2015: 11), are "more attuned to the relational and affective dynamics of narcissism, demanding not an attenuation of the subject's attachment to money but precisely an intensified and more authentic commitment, a spiritual purification to the subject's relation to the market."

In this respect, the idea of self-care becomes inexorably linked to material and spiritual consumption, as the ethos of "treat yourself became a capitalist command". Not surprisingly, in this respect, it is the age of market-friendly "self-help" gurus and wisdom (Illouz 2007). It is also telling that popular culture is now almost obsessed with the capitalist anti-hero, ranging from Tony Soprano to Heisenberg in *Breaking Bad* (see Tie 2004) to even Donald Trump as the irreverent US president. Together, these trends reflect the fervent effort to achieve free market righteousness and the salacious temptation of market-friendly sinfulness.

The religious transformation of capitalism is profoundly reconfiguring the theory and practice of market freedom. It increasingly finds itself identical to traditional understandings of Christian freedom. At stake is the emancipation from sin through strict self-discipline. Present is the liberty to practice your faith and the right to preserve orthodoxy in the face of heretical challenges. It is freedom to submit to a salvationary god, whether it be Jesus or the free market. Revealed, in turn, is the bad faith in the free market and the growing need for a more existential form of freedom.

An Existential Critique of the Market

The free market has been under increasing scrutiny and even outright attack. It is often pointed to as a failed system representative of a corrupt status quo. Its more radical critics lambast it as "voodoo economics", a mystical ideology that is out of touch with people's living realities. The demand from Liberals and progressives for the secularization of politics includes the religiosity of the free market. Yet amidst these various recriminations there is still little fundamental discussion of a need for a different theoretical and empirical account of freedom alternative to the one put forward by capitalism.

However, it is exactly, in this respect, that existentialism remains so potent and radical. It holds the promise of reactivating human agency, the ability of people—both individually and collectively—to reinterpret their realities and freely choose different ones. Such freedom is not naïve, blissfully ignoring of material conditions and psychic attachments. Instead it is a profound form of radical, critical reflection and action, challenging and even demanding subjects reject their idols and meaningfully engage with the shaping of their social condition.

Importantly, such existentialism transcends specific debates over the attractiveness of one set of freedoms over another. Conversely, it highlights the eternally unfinished business of freedom. Any attempt to naturalize freedom is questioned as necessarily precluding the deeper human ability to choose one's own destiny and way of life. Paradoxically, it is—as will be shown throughout this book—the intrinsic presence of freedom in our existence that undermines all attempts at making inherent any singular expression of what freedom is.

The free market, in turn, is challenged exactly based on its assertion that it has exhausted the possibilities of freedom. In doing so it runs up against a fundamental existential freedom. It is the fact that people ultimately have to choose this existence that its hegemony is only ever dependent on people literally and figuratively buying into it. The constant refrain that "There is no alternative" masks a contingent reality, where a historically specific form of Being is never so stable or indeed permanent.

Nevertheless, the chains of the free market are not simply broken by imagining them away. They are kept in place by our material and psychic attachment to our existing freedoms. While we may dismiss, in principle, capitalism, it is considered an incontrovertible fact that for us to experience any sense of agency we must abide by its prerogatives and expectations. To wit, one may not like wage labour, but there are few other viable avenues to limitedly shape our personal circumstances then by making oneself more employable in a competitive job market. It is all too easy, therefore, to dismiss the existential free choice as blind to our material realities or condemn those who choose the freedom at hand over one that does not yet exist. A chief aim of this book is to show how it is this gap between desire and ability, the longing for a different freedom and a present which makes it seemingly impossible to realize, that is fertile grounds for reactivating this fundamental and radical existential freedom—both in thought and action.

The existential critique of the market then is precisely on the grounds of freedom. While certainly concerned with its glaring economic and social problems, from inequality to global poverty, the thrust of its attack is that the free market limits our possibilities for conceiving and engaging with a diverse array of freedoms. Its critique is simultaneously eternal and always historically specific. It calls upon us to take an existential leap of faith beyond our current foundations, letting go of our limited freedoms, for the possibilities of experimenting with new ones still to be discovered.

Living in a "Post-Freedom" World

If existentialism promotes freedom as the foundation of all human being (and therefore beings), it also unsettles all existing freedoms. It is thus a contradictory force—at once a destroyer and creator of freedom. It troubles human existence as incomplete, universally pointing to that which it is not—its "non-being"—as a catalyst for transcending what is. However, it must then be asked, if human existence has no inherent meaning, is any freedom worth pursuing and investing in? Or are we condemned to an impossible freedom, more frustrating than liberating?

Market freedom would on the surface seem to offer an imperfect but plausible way out of this existential conundrum. While clearly not sufficient, it serves as a foundation for making "free choices" and pursuing a diversity of different types of existences. Yet, as shown, it ultimately subsumes all freedom to its economic and social demands, forcing them to submit with the fervour of a true believer to its quasi-religious beliefs. Consequently, it

blends the hardheaded approach to human capital of any successful firm with a national-theological discourse of moralized sacrifice, a sacrifice required for the health and survival of the whole. Moralized sacrifice finesses the paradox of unrewarded conduct normatively prescribed by neoliberalism. (Brown 2015: 4)

The answer lies, hence, not in the total investment in refounding freedom. By contrast, it is in the ironic acceptance and radical engagement with our fundamental lack of any foundations. In the first instance, this implies recognizing that life is in fact meaningful—that it is filled with human created meanings that are available for reinterpretation. It also requires an honest assessment and judgement as to which aspects of our existence are we meaningfully prioritizing and in turn what sort of social reality does this serve to reinforce and reproduce?

However, it also sets a new challenge for humans in relation to freedom. It is not so much whether we embody or have perfected a prevailing version of freedom but instead our willingness and capacity to construct new freedoms. This challenge is, of course, multifaceted. Specifically, it calls upon us to set ourselves paradoxically free of freedoms—to recognize which forms of agency are currently defining our existence and choosing to subvert and potentially upend them. Yet, this admittedly ironic theoretical troubling of freedom—regardless of how abstractly radical—is itself insufficient. It is only the first step. The next is to assess and instantiate new forms of freedom liberated from those of the past (though not necessarily completely disregarding of them).

Doing so, obviously, is easier written about then done. What is required is the commitment to a philosophical perspective that constructively engages with our foundationless existence. While often dismissed as overly esoteric or an untenable basis for practically theorizing freedom—current post-structuralist thinking offers just such an opportunity. There is a common critique that post-structuralism is a complete rejection of social structures in general, the total relativization of meaning as such. Nevertheless, this is only part of its philosophical story. It is also a theoretical call to critically reflect on what is meaningful, which structures prevail, in order to better understand how they can be reinterpreted and how existing culture can operate differently.

Post-structuralism is thus an attempt, in all its broadness, to theoretically fulfil the existentialist insight that places existence before essence. It is not a denial that freedom is present or that certain meanings have value. Rather it is an acceptance that these will always be partial and contingent, respectively. Furthermore, it is perhaps the most full philosophical expression of Sartre's observation that we are "condemned to freedom"—that we must make free choices in a world in which we are rarely free to fully do so. Similarly, post-structuralism condemns us to acknowledge and take seriously that we as humans are the ultimate makers of our meaning and the creators of our living reality.

If there is an underlying pessimism to existentialism, a frustration with a freedom that is our birthright but rarely our living right, then there is a rather strange optimism to post-structuralism—especially when it comes to freedom. In recognizing and critically engaging with our own social construction, we become free to recreate our existence. To this end, it embraces existence by first deconstructing its perceived essence. Hence, by questioning market freedom, not only on normative grounds but on existential ones, interrogating its history, meanings, and associated contingent practices, we become partially emancipated from its hold over us.

Consequently, freedom moves from being almost exclusively defined by the market to a full-fledged creative enterprise. More precisely, through de-essentializing freedom generally and in relation to capitalism particularly, we become free, in turn, to creatively reimagine and materially experiment with what it could be. In the contemporary age, there is often a lament that we need more time and freedom to be creative. However, existential offers a more radical proposition—that we can be creative with our freedom, that existence is never pre-given and therefore always open to be changed.

What is key is to embrace the possibilities of a post-freedom world. Put differently, to critically view any freedom with a scepticism and as a possibility for reinvention. Freedom, in this respect, becomes not values to be enshrined and preserved—though undoubtedly their protection at certain times can certainly be desirable and necessary—but a constant existential jumping-off point to find out what lies beyond their limited social horizons. It is a post-free world not in the rejection of freedom but rather in the fact that through constant processes of interpretation its potentialities are never exhausted and its exciting possibilities always contained in Being to come.

Going Forward

This book attempts to do more than simply challenge market freedom. Instead, it desires to critically explore the market freedom produced by the free market and what opportunities they provide for the creation of radically new freedoms. Specifically, it explores how we can use critical theories from Marxism to ideology and discourse analysis to psychoanalytic fantasies to deconstruction in order to existentially reinterpret and transcend capitalism and the free market. The goal, in this respect, is to reject our bad faith in the free market and reinvest in a good faith over the fresh possibilities of freedom and existence.

Following this introduction, Chap. 2 will explore the liberating potential held by existentialism for contemporary efforts to move beyond restrictive and orthodox market freedoms. Drawing on Sartre's seminal early work "Existentialism and Freedom" it will highlight continuing importance of the insight that "existence proceeds essence", revealing the dangers of reifying human inspired and historically specific ways of understanding and living in the world. It will then trace out how the free market evolved from a radical existential promise for enhancing human freedom to a dogmatic discourse limiting its development and growth. Significantly, it will introduce the concept of an "existential gap" referring to the always existent chasm between our conscious capacity to choose how we interpret the world and our often material and social inability to substantially do so. Finally, it will highlight the positive existential gap opened up by market freedom, revealing a contemporary "existential challenge" to "break free from market freedoms" and construct "new foundations of freedom" in its place.

Chapter 3 will show that the 2008 financial crisis more than challenged the faith in the free market; it represented a collective existential crisis where people questioned whether capitalist society and life had meaning. Inspired by Sartre's later work in "Search of a Method", it will examine the evolution of freedoms from an empowering philosophy for remaking the world to a stifling discourse limiting such existential agency. It will then reveal the insights Marxism provides in terms of the socio-material production of freedom and the reframing of history as the revolutionary reproduction of progressive human freedoms. The near global financial meltdown thus posed the possibility for not only reforming or even radical transforming the free market but also breaking free from the "fundamentalism" that posited capitalism as the only possible option.

Chapter 4 explores how the "facts" of neoliberalism must be transformed into "facticities", conditions that are currently holding us back from realizing our existential freedom. It first counterposes supposed neoliberal "facts"—such as the need to be "fiscally responsible" or the idea that massive "inequality" is acceptable and even desirable—with Satre's interpretation of "facticities", the events, conditions, and capabilities that impact on what one can and cannot do. This reading allows for an understanding of how these dogmatic "facts" represent an ideological hegemony that forecloses the opportunity to conceive and practically explore alternative modes of freedom. This insight will be critically investigated, in particular, though using the theories of discursive hegemony first introduced by Laclau and Mouffe. The relabeling of neoliberal "facts" as "facticities" thus opens the potential for constructing a counter-hegemonic politics aimed at expanding a dominant social horizon of freedom, in this case away from the narrow limits of the free market.

Chapter 5 expands on the previous one by interrogating the deeper psychic relationship between capitalism and nothingness. Taking its cue from Sartre's famous philosophical text "Being and Nothingness", it contends that capitalist existence is built on positing a continual nothingness—or sense of experienced lack—that only the market can fulfil. In times of crisis, this turns apocalyptic with capitalism being posited as the only thing that can prevent total nothingness. At the affective heart of the embrace of this market "unfreedom" is the underlying fear that without capitalism we would dissolve into nothingness, a worry captured in the notion of the "real" first put forward by the renowned psychoanalytic thinker Jacques Lacan—our fragmentary "true" nature that must be masked by a comforting fantastic "reality". It concludes by positing the need for creating a radical fantasy of freedom that embraces this nothingness as part of a fundamental drive to be eternally dissatisfied with hegemonic forms of freedom such as those linked to capitalism and therefore seeks out new ones.

Chapter 6 investigates how capitalism turns individuals into "subject-objects", combining our economic objectification to market demands with the possibility to choose from a range of market-friendly identities. It begins by highlighting Sartre's belief presented in Part II of "Being and Nothingness" that individuals are denied their existential freedom by turning themselves into socialized objects whose only purpose is to fulfil their given cultural role. The French theorist Foucault, for his part, notes how this subjection to being a mere object of capitalism (or any social system) is offset and ultimately reinforced through processes of subjectification in which we are made into conscious and intentional subjects. A new philosophy of existential freedom would, hence, emphasize the overriding importance of reducing this alienation, providing us the material resources and subjective tools to freely produce our own selves and society.

Chapter 7 explores how the personal freedom often associated with the market and existentialism can be combined with a radical collective freedom. It continues from the previous chapter by highlighting how for Sartre there is always a fraught tension between our radical existential freedom and our "identity projection", that which we assume ourselves to be based our social roles and expectations. Thus, for Sartre, existential freedom is most concretely experienced through individuals forming their own continual "life project" that focuses not on what one is but has not yet become. Capitalism, of course, hints at just this possibility in its obsession with individual social mobility and the liberating potentials of market rationality yet is limited by its orthodox commitment to market values. However, existential holds out an eternal "promise"—drawing on the ethico-political philosophy of Jacques Derrida—for enacting a never perfected but also perfectable freedom. In this respect, existential freedom hangs like a spectre over any and all dogmatic systems—including the present-day free market.

Chapter 8 will summarize the present dangers of our continued bad faith in the free market. It would argue that all of the theorists covered, moreover, suffer from their own implicit bad faith. Marxism in its dogmatic commitment to class struggle and revolution, Foucault to the inherent "danger" in all movements and experiences of freedom, and Lacan to the impossibility of ever truly overcoming our fundamental psychic lack. Yet together they pose a compelling and inspiring vision of freedom. It will end optimistically by showing how this combining of existentialism, Marxism, post-structuralism, and psychoanalysis can open up the radical possibilities of having good faith in our potential for subjectively and materially becoming existentially free.

References

Brown, W. (2015). *Undoing the Demos*. Boston: MIT Press.

Drèze, J., & Sen, A. (2002). *India, Development, and Participation*. Oxford: Oxford University Press.

Foner, E. (1999). *The Story of American Freedom*. Winnipeg: Media Production Services Unit, Manitoba Education.

Friedman, M. (1970). The Social Responsibility of Business Is to Increase Its Profits. *The New York Times*.

Furnham, A., Bond, M., Heaven, P., Hilton, D., Lobel, T., Masters, J., et al. (1993). A Comparison of Protestant Work Ethic Beliefs in Thirteen Nations. *The Journal of Social Psychology, 133*(2), 185–197. https://doi.org/10.1080/0 0224545.1993.9712136.

Giorgi, L., & Marsh, C. (1990). The Protestant Work Ethic as a Cultural Phenomenon. *European Journal of Social Psychology, 20*(6), 499–517. https://doi.org/10.1002/ejsp.2420200605.

Illouz, E. (2007). *Cold Intimacies: The Making of Emotional Capitalism*. Cambridge: Polity Press.

Jameson, F. (2003). Future City. *New Left Review, 21*, 65–80.

Jenkins, K. (2001). *On "What Is History?"*. London: Routledge.

Konings, M. (2015). *The Emotional Logic of Capitalism: What Progressives Have Missed*. Stanford, CA: Stanford University Press.

Kozul-Wright, R., & Rayment, P. (2007). *The Resistible Rise of Market Fundamentalism*. London: Zed.

MacGilvray, E. (2011). *The Invention of Market Freedom*. Cambridge: Cambridge University Press.

Sen, A. (1985). Well-Being, Agency and Freedom: The Dewey Lectures 1984. *The Journal of Philosophy, 82*(4), 169. https://doi.org/10.2307/2026184.

Stiglitz, D. (2009). Moving Beyond Market Fundamentalism to a More Balanced Economy. *Annals of Public and Cooperative Economics, 80*(3), 345–360. https://doi.org/10.1111/j.1467-8292.2009.00389.x.

Tie, W. (2004). The Psychic Life of Governmentality. *Culture, Theory and Critique, 45*(2), 161–176. https://doi.org/10.1080/1473578042000283844.

Weber, M. (2002). *The Protestant Ethic and the Spirit of Capitalism*. United States: Wilder Publications.

2

Breaking Free from the Free Market: The Existential Gap of Freedom

There is a growing belief that, despite its claims, the free market is in point of fact not especially free. The clear evidence of its negative effects such as rising inequality and economic insecurity makes this conclusion seem even more obvious. Yet this combined empirical and normative critique should be a catalyst for critically reflecting on contemporary freedom more generally. In other words, it must lead to a positive reimagining of current conceptions and enactions of liberty and emancipation. More precisely, at its most radical and profound, these critiques can serve as an opening for theoretically exploring the scope of present freedom as well as how these efforts are being practically impeded by contemporary capitalism.

As the first chapter explained in greater depth, a key claim for the legitimacy of market freedom is that it is the exclusive or at least the best pathway for the realization of individual freedom. Recent times have witnessed this relatively straightforward equation completely reversed. The market is posited by a growing many as a barrier to feeling free. It is an oppressive force that blocks an individual or a community from fulfilling their potential. It detracts from the supposedly inherent right to "life, liberty, and the pursuit of the happiness". On a purely material

© The Author(s) 2018
P. Bloom, *The Bad Faith in the Free Market*,
https://doi.org/10.1007/978-3-319-76502-0_2

level, the precariousness of simply being able to materially persist makes it a capitalist reality more of a struggle for survival than an exercise in human flourishing and freedom.

Yet even when material survival is secured, the social privileged and economically well paid, their lives are often defined by a slavish devotion to profit and the trappings of success. However, they are also marked by what Rachel Sherman (2017: 2) has recently referred to as the "anxieties of affluence" in which elites face "conflicts about how to be both wealthy and morally worthy, especially at a moment of extreme and increasingly salient economic inequality". Significantly, these elites have only a minimal amount of ability—or incentive—to fundamentally alter the system that they ostensibly benefit from than those that are so clearly oppressed by it. The tyranny of the free market, hence, is simultaneously quantitative and qualitative in nature.

The modern-day critiques of the free market bear a striking resemblance to the revolutionary anti-capitalist discourses of the late nineteenth and early twentieth century. Indeed there has been a veritable reawakening of left-wing thought and politics. From the so-called "Arab Spring" to the Momentum movement in the UK to "our Revolution" in the US inspired by the surprising success of "democratic socialist" Bernie Sanders's presidential campaign. Each shares an explicit condemnation of modern capitalism, raging against its growing economic inequality, entrenched racism, and the global "race to the bottom". Indeed

> Support for anti-establishment parties in the developed world is at the highest level since the 1930s-and growing … The reasons for this backlash are rather obvious. The financial crises of 2007–2009 laid bare the scorched earth left behind by neoliberalism, in which the elites had gone to great lengths to conceal in both material (financialization) and ideological ("the end of history") terms. (Mitchell and Fazi 2017: 1)

However, these diverse movements also have in common a resurgent radical desire for freedom. Here the very justification for capitalism and the present-day promotion of is neoliberalism is almost completely reversed. Now, it is the crucial task is to be free of the "free market". Emancipation for an increasingly desperate and indebted population is found in breaking the socio-economic chains of a predatory financial system and its elite profiteers. As even an article in the traditionally rather Centrist *New Statesman* proclaimed in 2017,

> The 2008 financial crisis led to an enduring loss of faith in economic elites. Capitalism has since failed to deliver on its promise of rising living standards for the majority. And voters have grown ever more weary of public spending cuts. (Eaton 2017: N.P.)

The desire for freedom has turned decidedly against the free market. There is a gap, a break in history, where a once assured source of freedom is transformed into its greatest nemesis. What this new freedom is, of course, remains to be discovered. Yet even in such an infant state it haunts the status quo as an invisible but unsettling ghost. In the movement of the present there is a vibrant challenge to the current limits on liberty. A political spark building upwards to a potential firestorm based on the belief that the possibilities of freedom have not been exhausted and the potential of the human spirit to shape its development has not been extinguished. It is nothing less than the beginning of a mass existential reawakening.

The Growth of the Unfree Market

The free market arose based on its romanticized promise of freedom. Indeed the very utterance of its name—the free market—conjures up images of being liberated from social constraints to live freely. Here the enemy is big government, regulations, and the tyranny of bureaucracy. Trust is placed in the market to cut through the red tape tying up and holding down all our aspirations. In practice, the free market has evolved into a similarly dogmatic and repressive social system.

The history of the free market is the conventional story of a revolution corrupted. Marx famously wrote that history happens twice "the first time as tragedy, the second time as farce". So too is the case of modern capitalism it would appear. It began as an economic ideology meant to emancipate people from deprivation and mercantilistic economies that primarily benefited the socially privileged classes. The industrial revolution has its roots in the eighteenth-century political and philosophical upheavals championing meritocracy and the rights of individuals to "life, liberty, and the pursuit of happiness". They reflected an "enlightenment narrative" chronicling the historical

> descent from classical antiquity into the 'barbarism and religion', and the emergence from the latter set of conditions of a 'Europe' in which civil society could defend itself against disruption by either. This history had two themes: the emergence of a system of sovereign states ... and the emergence of a shared civilization of manners and commerce. (Pocock 2001: 20)

These radical beginnings soon, however, gave way to the replacement of one set of elites for another—as capitalists overtook aristocrats and royals in both influence and increasingly political power. More to the point, it transformed into a system whose logic was less concerned with emancipation as it was with global conquest. Consequently,

> the empires that mobilized this first European expansion ... also obstinately refused to imagine forms of social coexistence not ruled by the logic of possession, consumption, commoditization, and violence. Colonizing from its inception, this first modernity built itself upon a structure of political, economic, and theological power that claimed universal applicability, and rendered any expression of difference invisible or subaltern. (Monasterios 2018: 553)

Crucially, capitalism has always represented an ideal as much as a reality. Put differently, it exists as both a regulatory material system and a utopian like dream of a better market world. It certainly, in this regard, created an "active" worldview for making social sense of the world. However, it has been just as powerful as an idealized critique against tyrannies, advocating for the sanctity of human freedom generally. It championed the ability of people to control their own destinies through their own hard work and talents and innate talents. It followed a similar historical logic to the supposedly autonomous and self-created rise of Europe to economic and political dominance—one that masked its reliance on the exploitation of others as well as the variety of contextual factors that made their supposed "superiority" possible. Hence

the internalist story of an autonomous and endogenous 'rise of the west' constitutes the founding myth of Eurocentrism. By positing a strong 'inside-out' model of social causality (or methodological internalism)— whereby European development is conceptualized as endogenous and self-propelling—Europe is conceived as the permanent 'core' and 'prime mover' of history. (Anievas and Nişancıoğlu 2015: 7)

Nevertheless, by the end of nineteenth century, this supposedly liberating ideology had evolved into a full-scale oppressive idealization. More precisely, what once was an expansion of human agency had become a restraining orthodoxy limiting its possibilities (see Appleby et al. 1996). In this respect, "freedom meant prosperity; freedom meant progress; freedom meant having willing workers as opposed to unwilling ones" (Temperley 1977: 109). It was the gilded age where the gold at the top could not cover the mass material deprivation and inequality. It required, therefore, "the expansion of bureaucratic states as power structures maintaining police and military control over potentially rebellious populations and reproducing the conditions of capitalist accumulation" (Alford and Friedland 2011: xiii). Further, previously promising democracies were increasingly bought and sold to the highest bidder. The emergence of Marxism and socialism as one of the defining philosophies of the twentieth century was inexorably linked to its urgent critique of a capitalism that had run amok. In the famous words of Marx and Engels "A spectre is haunting Europe—the spectre of communism" (2016: 9).

A century later the tables had turned considerably. Following the carnage of WWII and the global misery of the great depression before that, it was taken almost as a matter of course that the market needed substantial regulation. The question was no longer if the state should intervene in the economy but rather to what extent (see Kavanagh 1987; Leys 1997). The initial stirrings of the "free market" were themselves born out of their own critical stance to this so-called post-war "liberal consensus":

With modernization discredited and no single overriding narrative of progress to replace it, neoliberals took the field with their own promises of accelerated, benevolent change … Neoliberalism, in other words, prevailed precisely because it revived a vision of the global mission of the United States and made the same sort of transformative claims that modernization had. (Latham 2010: 158)

Thus, it was in its own admittedly Conservative manner a revolutionary challenge to a new world order defined by the struggle between Liberalism and Communism.

Tellingly, the movement that would later be both academically and, to an extent, popularly referred to as neoliberalism arose as a clarion call for freedom. It championed the nurturing of a market freedom that was both intrinsic and instrumental in character—equally inherent and practical in its provision of human liberty. Its proponent, to this end, argued that it reflected nothing less than the most pure expression of natural human freedom (Bernstein 1971; Tipps 1973).

The establishment of the free market permitted the unrestrained pursuit of personal interest as well as the supposed most assured route to avoid the bondage of the past. One of the most influential and renowned neoliberal thinkers Hayek writes in his book *The Road to Serfdom* that

when economic power is centralized as an instrument of political power it creates a degree of dependence scarcely distinguishable from slavery. It has been well said that, in a country where the sole employer is the state, opposition means death by slow starvation.

It was deeply, therefore, existential in its attraction. The coming of the free market promised people the potential to create their own meanings and be firmly in control of their own individual existences. Its later mass appeal as the catalyst for the creation of an "aspirational society" is hence completely understandable (despite as would be borne out by experience largely misguided).

There is an often all too common tendency when analysing neoliberalism to reduce it to its crudest economic level—pure greed and self-interest. Doing so ignores the profoundly emancipatory heart of neoliberalism, the market as the supposed liberation of all people from the clutches of government bureaucracy to follow their dreams. It was a symbolic stroke against the threat of an all-controlling party or Big Brother, for a new free market system in which individual possibility was in principle limitless.

Ironically, the rise of the free market was also one of the most sustained and politically successful examples of a collective movement for greater existential freedom. From its roots in the 1950s as a marginalized economic critique to its development into a radical right-wing force in the 1960s to an increasingly mainstream populism in the 1970s, it grew into power on a wave of fresh enthusiasm that together humans could dramatically transform their social condition. It was promoted as nothing less than a full-scale "neoconservative revolution" declaring that "the state must never govern society, dictate to free individuals how to dispose of their private property, regulate a free market economy or interfere with the God-given right to make profits and amass personal wealth" (Hall 2011). It demanded not mere change or tinkering around the edges of a seemingly adrift and stagnating economic order but a full-scale alteration of its principles and practices.

This sunny revolutionary optimism was most welcomed by many within a population caught in the apparently inescapable battle between of Liberalism that had lost its way and a really existing socialism that had become the modern symbol of tyranny. Nevertheless, the destructive consequences of this "new dawn" were quick to appear and ultimately longstanding. Inequality skyrocketed, poverty increased, civil liberties were curtailed as historically marginalized groups were further demonized and repressed. Required fundamentally was expansive "police and legal structures and functions required to secure private property rights and to guarantee, by force if need be, the proper functioning of markets" (Harvey 2005: 2). Moreover, the two largest original proponents of neoliberalism—the US and the UK—were propped up economically from massive public defence spending and the discovery of Scottish oil, respectively.

In the face of mounting empirical evidence that its freedom was a mirage, its supporters turned to touting its inherent necessity. These sentiments followed in a tragic modernist tradition of cloaking contingent political ideas in supposedly unassailable rationalist dogma. As such

Neoliberalization has in effect swept across the world like a vast tidal wave of institutional reform and discursive adjustment. While plenty of evidence shows its uneven geographical development, no place can claim total immunity (with the exception of a few states such as North Korea). Furthermore, the rules of engagement now established through the WTO (governing international trade) and by the IMF (governing international finance) instantiate neoliberalism as a global set of rules. All states that sign on to the WTO and the IMF (and who can afford not to?) agree to abide (albeit with a "grace period" to permit smooth adjustment) by these rules or face severe penalties. (Harvey 2005: 23)

The free market was proffered as absolutely essential for the development of liberal democracy politically. Reasoning was in time evolved to reflect the idea that these policies, whether or not popularly desired, were based on iron-clad economic laws.

By the end of the twentieth century and start of the new millennium, it was simply accepted that the free market was a necessary and unchangeable reality that could not be fundamentally altered. At best it could be politically negotiated with in regards to the terms and relative limits of its overall social and economic domination. Human freedom was once again reduced to small-scale battles over the fine print of an entrenched and permanent form of existence.

Existentialism and Humanism

The rise and stagnation of the free market has, as shown, come at a great potential cost to human freedom. In championing market freedom, it robbed individuals and communities of the ability to define for themselves the world and act accordingly. Rather, it offered them a pre-packed bill of goods selling not only the benefits but also the immutability of this capitalist agency. Thus ironically in promoting its own freedom, it suppresses humanity's more fundamental freedom. This paradoxical championing and repression of freedom reflects the continuing philosophical and practical significance of existentialism for our times.

While existentialism is by no means defined by any one thinker or set of ideas, arguably the most notable and comprehensive of its proponents was the French philosopher Jean-Paul Sartre. One of the most renowned and influential thinkers of the twentieth century, during his long, illustrious, and at times controversial career a philosopher, playwright, novelist, political theorist, and even biographer. His best-known works include the philosophical tome "Being and Nothingness", his novel *Nausea*, his plays such as *No Exist*, and his later unfinished philosophical treatise seeking to combine existentialism and Marxism *Critique of Dialectical Reason*. Both a radical and widely admired thinker of his age, Sartre famously refused the Nobel Prize for literature in the 1960s stating he did not want to be "transformed" by the award and was pardoned from arrest for his role in the 1968 Paris protests by none other than President Charles De Gaulle who was quoted as saying "You do not arrest Voltaire".

One of his earliest and most widely cited philosophical defences of existentialism was in his 1946 published lecture "Existentialism and Humanism". Though he would later distance himself from many of its key claims, it remains a compelling place to begin exploring the contemporary relevance of existentialism. In it, he reaffirms a central premise of his longer earlier work "Being and Nothingness" that "existence precedes essence". Specifically, he refers to the fact that there is no creator guiding our actions nor external force predetermining them. Therefore, we are "condemned to freedom" as we are tasked with recognizing that we are their progenitor and consequently must take full responsibility for them. Quoting Sartre (1948: 33–34) at length,

> existentialist, on the contrary, finds it extremely embarrassing that God does not exist, for there disappears with Him all possibility of finding values in an intelligible heaven. ...For if indeed existence precedes essence, one will never be able to explain one's action by reference to a given and specific human nature; in other words, there is no determinism—man is free, man is freedom. Nor, on the other hand, if God does not exist, are we provided with any values or commands that could legitimize our behavior. Thus we have neither behind us, nor before us in a luminous realm of values, any means of justification or excuse. We are left alone, without excuse. That is what I mean when I say that man is condemned to be free.

Condemned, because he did not create himself, yet is nevertheless at liberty, and from the moment that he is thrown into this world he is responsible for everything he does…

While an ultimately optimistic and potentially liberating philosophical account of human existence, Sartre admits that this realization of our fundamental freedom produces a sense of anguish. More precisely, it is an emotional response at coming to face with their freedom and the responsibilities this entails. According to Sartre (Ibid.: 32),

In making the decision, he cannot but feel a certain anguish. All leaders know that anguish. It does not prevent their acting, on the contrary it is the very condition of their action, for the action presupposes that there is a plurality of possibilities, and in choosing 4 one of these, they realize that it has value only because it is chosen. Now it is anguish of that kind which existentialism describes, and moreover, as we shall see, makes explicit through direct responsibility towards other men who are concerned. Far from being a screen which could separate us from action, it is a condition of action itself.

However, it is also exactly this anguish that in his view propels us to make judgements as to how we would like to interpret and live in the world, a judgement that extends not only to ourselves but to humanity generally.

In this spirit, he introduces his strangely hopeful concept of existential despair. Far from its usual connotations of abjection, it is an acceptance of oneself as the free shaper of their lives. He famously declares, in this respect, "In fashioning myself, I fashion Man." It is this despair that serves as a catalyst for individuals to embrace their conscious existence as a "being-for-itself" rather than an unconscious and completely naturalized "being-in-itself". This explicit realization of their being allows them to actively engage with their freedom, make free choices, and accept their consequences. Thus

Its intention is not in the least that of plunging men into despair. And if by despair one means—as the Christians do—any attitude of unbelief, the despair of the existentialists is something different … Not that we believe God does exist, but we think that the real problem is not that of

His existence; what man needs is to find himself again and to understand that nothing can save him from himself, not even a valid proof of the existence of God. In this sense existentialism is optimistic. It is a doctrine of action, and it is only by self-deception, by confining their own despair with ours that Christians can describe us as without hope. (Ibid.: 56)

Sartre does though acknowledge the darker aspects of this freedom. In particular, he introduces the notion of abandonment to depict the loneliness people experience when they must confront that we are alone in the universe—without a god or preordained nature to guide our beliefs or practices. Hence, "That is what 'abandonment' implies, that we ourselves decide our being. And with this abandonment goes anguish" (Ibid.: 39). We are thus consigned to a sense of divine abandonment and an acceptance that we are the ultimate determiners of our own fate. Whilst this certainly can be emotionally difficult, it is also a necessary trauma to the empowering embrace of our freedom.

"Existentialism and Humanism" therefore provides one of Sartre's earliest and most passionate testaments to the potentially liberating implications of existentialism. It proposes a human existence which at its essence is defined by its freedom. Although propounded over half a century ago, this account of radical freedom resonates with the current epoch still steeped in our dogmatic acceptance of the free market. It stands as a continuing clarion call to reclaim our freedom by having the courage and will to topple our false market idols.

The Gap of Freedom

Existentialism may appear to offer a rather straightforward account of freedom. We are the makers of our own reality and thus must accept the responsibility that this implies. It is not so much whether we want to be free, in Sartre's view. Rather it is that we are free and it is our choice whether to accept it or not. However, this does raise critical complications for the theory and practice of freedom. Specifically, it highlights an existence that is simultaneously already free and always striving to be freer.

A common refrain from the past and present is the desire "to be free". This sentiment signifies the longing to shed oppressive norms and power relations inhibiting individual agency. Despite being close to a cliché, it would be seemingly hard to argue such aspirations. However, existentialism to an extent reverses this conventional formula—it asks how can we be free of Being. To this end, a chief component of being free is precisely the recognition that there is the possibility of existing beyond the present order. That the possibilities of Being are never exhausted and as such freedom requires thinking and moving beyond its current version.

Introduced then is a crucial paradox of freedom. On the one hand, freedom is fundamental to human existence. There is no God or underlying transcendental force dictating our actions or inherently structuring our experience of reality. Our experiences are never predetermined or predestined. They are ours to shape as we so choose. On the other, every moment is a further realization that such freedom remains incomplete. The very act of freedom is grounded in the consciousness that one is not yet totally free.

Freedom thus resides in this tension between these competing modes of being. The German philosopher Martin Heidegger—a major influence on Sartre—introduces the ontological difference distinguishing between the general structure of Being and the actual existence of beings. Being gives birth to beings but is never exhausted by them nor is Being ever fully revealed by the actual experiences of being. Similarly, taking inspiration from Sartre's insights, it can be said that there is a freedom difference, in so much that Being implies total freedom and yet being and beings are only ever at best partially free. Freedom is then never a finished product.

Consequently, freedom is at once a liberating promise and an inscriptive reality. It is an eternally elusive birthright, the cornerstone of human existence shadowing each and every one of our decisions. However, it additionally exists as a concrete means for shaping our reality and avoiding being completely defined by our environment. In this respect, freedom constantly runs the risk of being essentialized. We rarely if ever experience pure freedom, it is always a limited version of it. Moreover, this partial expression of freedom can easily become reified and all-pervasive— put forward as the one and only way to experience a sense of agency full stop. And as is the case with market freedoms, these socialized freedoms are commonly justified as indicative of our deeper "human nature".

There is therefore an existential gap at the core of our existence. Namely, it is the chasm between our longing to be totally free and our recognized actuality that we are not so. This gap is constantly being filled by social discourses trumpeting specific types of freedoms. Hence, freedom evolves into the very thing which it is meant to destroy—an essentialized force for determining human existence. It is by breaking free, ironically, from existing freedoms that that gap of freedom is widened enough to allow new freedoms to exist.

The Present Challenge of Existential Freedom

The global growth of the unfree market has largely defined the twenty-first century. The intentional spread of neoliberalism to all corners of the world reflect less a liberation for oppressed populations and more the acceptance of a repressive system that can neither be stopped or fundamentally altered. As noted political theorist Colin Crouch (2012: N.P.) observed,

> Many fear that neoliberalism will never be defeated. They may be right if their fears are that the interests sustaining the neoliberal system are too powerful. When they claim neoliberalism will prevail because there are no viable alternatives, however, they are quite wrong. The ideas are out there; they are widely understood and coherent; there are even good examples of them in action.

Its seemingly inevitable reach extended beyond geographic boundaries, ceaselessly expanding with an unstoppable certainty into all areas of cultural existence.

While there is increasing emphasis placed on the material effects of this total marketization, its negative contribution to our shared freedom has received considerably less attention. However, it is becoming increasingly clear that neoliberalism poses perhaps above all else a profound modern-day existential challenge. Thus

The conceptual expansion of neoliberalism from economic policy to political power is present more broadly in the literature as the idea of 'capital resurgent': a reassertion of capitalist class power that seeks to disengineer the post-war compromises of tripartite corporatism and expanded social welfare. (Venugopal 2015: 168)

Highlighted in turn was a dramatic reversal to the public perception of capitalism generally and the free market specifically. Traditionally their desirability was intimately related to their providing a compelling supposed answer to the fundamental human question of freedom. Hence all non-market systems were customarily rejected on principle as being inherently unfree. Consequently, even while acknowledging that it was not working perfectly, British Prime Minister Theresa May publicly stated that the free market remained the "only sustainable means of raising the living standards of everyone in a country" (Elliott 2017: N.P.).

Theoretically, this points to post-foundationalist ideas of "problematization". First popularized by Foucault, it refers to the ways a specific social concern or issue comes to predominantly shape existent social knowledge and practices. It denotes "[not] behaviours or ideas, not societies and their 'ideologies' but the problematization through being offers itself to be, necessarily thought—and the practices on the basis of which these problematizations are formed" (Foucault 1985: 11–12). This problematization takes on an existential quality in its focusing of individuals on specific freedoms at the expense of others, thus limiting their overall agency for reinterpreting and transforming their existence. Hence, according to the renowned French Marxist Louis Althusser (1969: 164),

[T]o say that this is a problem implies that we are not dealing merely with some imaginary difficulty, but with a really existing difficulty poses us in the form of a problem, that is in a form governed by imperative conditions.

In this respect, the free market is exclusively focused on realizing market freedoms, serving as discursive and practical barrier for the exploration of alternative ways of seeing and being in the world.

The failure of neoliberalism to deliver on this promise of freedom, however, has catalysed a new opportunity to redefine and engage with freedom. An all too common lament of the contemporary age is that the attacks against the reigning status quo with equal passion from the perceived margins of both the Right and Left. For those from the privileged "centre ground" they may appear to be nearly identical barbarians threatening at the gates of their free market civilization. In the words of social commentator Pankaj Mishra (2016: N.P.), "The seismic events of 2016 have revealed a world in chaos—and one that old ideas of liberal rationalism can no longer explain". While such political myopia is an obvious indication of elite blinders, the anti-establishment ethos growing across the ideological spectrum, nevertheless, reflects a shared existential frustration. It gestures towards a rising mass desire for people not dogmatic ideologies or social systems to determine their own social destiny.

At the heart of these movements is a beating desire for recapturing a personal and collective sense of existential freedom—even if obviously very few if any would articulate it in such explicit terms. The defining feature of the free market is no longer its emancipatory possibilities—its trumped-up claims of limitless individual mobility and liberation from a tyrannical state. Rather it is found in its perceived inevitability. Indeed,

> This populist backlash reminds us that the rewards of globalization are not evenly distributed, and as a result there has been some questioning of the idea that borders should be open to trade—as well as concerns about what might happen instead. (Ghemawat 2017: N.P.)

The discourses surrounding globalization stand as a prime contemporary example of this supposed inescapable limit imposed by the market on human potential.

What is being witnessed, hence, is the reintroduction of existential freedom as a driving political force. It has been reactivated as an urgent demand for freedom upon the status quo. Theoretically, this can be described as the transformation of the problem of freedom into the challenge of freedom. Reflected is the shift in sentiment from trying to merely perfect an existing form of freedom to demanding, even if initially only as a form of critique, a renewal of human agency to shape the present and future. Freedom, in turn, goes from a problem to be solved to an intervening challenge calling for radical solutions.

Breaking Free from the Unfree Market

The free market is increasingly assailed for its creation of present material inequities and over fears that it will be responsible for our future material destruction. Critically, it is condemned for its creeping corruption of our democracy and civic society. To this end,

> Both persons and states are construed on the model of the contemporary firm, both persons and firms are expected to comport themselves in ways that maximize their capital value in the present and enhance their future value, and both persons and states do so through practices of entrepreneurialism, self-investment, and/or attracting investors. (Brown 2017: 22)

Explicitly relevant to questions of freedom, scholars are progressively linking the rise of marketization with the strengthening of political authoritarianism and civic illiberalism (see Bloom 2016). However, the free market also is open to profound criticism on existential grounds. A critical aspect of this system, in this regard, is its universal promotion of market freedom as necessary and desirable in all areas of human existence. This spread of marketization extends, as mentioned above, from the political realm all the way to our interpersonal relationships. Significantly, its appeal rests in its social framing as the best means for experiencing freedom regardless of one's aspirations. It stands as the very "foundation of freedom" according to Eamonn Butler (2013: 13) as "Freedom creates prosperity. It unleashes human talent, invention and innovation, creating wealth where none existed before. Societies that have embraced freedom have made themselves rich. Those that have not have remained poor." It is sold to the public as an essentialized freedom, an inherent means for personally shaping their reality to reflect their diverse desires. Tellingly, it

> has become incorporated into the common sense way many of us interpret, live in, and understand the world. The creation of this neoliberal system has entailed much 'creative destruction', not only of prior institutional frameworks and powers … but also of divisions of labour, social relations, welfare provisions, technological mixes, ways of life and thought, reproductive activities, attachments to the land and habits of the heart. (Harvey 2005: 3)

Nevertheless, this legitimization of the free market has recently taken a rather noticeable existential turn. In the wake of a financial crisis, the religiosity surrounding market freedom has somewhat waned. As such

> Once the failure of free trade, deregulation, and monetarism came to be seen as leading to a "new normal" of permanent austerity and diminished expectations, rather than just to a temporary banking crisis, the inequalities, job losses, and cultural dislocations of the pre-crisis period could no longer be legitimized—just as the extortionate taxes of the 1950s and 1960s lost their legitimacy in the stagflation of the 1970s. If we are witnessing this kind of transformation, then piecemeal reformers who try to address specific grievances about immigration, trade, or income inequality will lose out to radical politicians who challenge the entire system. And, in some ways, the radicals will be right. (Kaletsky 2017)

In its place have come renewed philosophical questions as to whether it in does represent "human nature" and practically whether it is the only way forward. Revealed is a growing anguish with neoliberalism, a distancing of ourselves from its dogmatic embrace and a realization that we are free to determine which norms and practices should guide our existence.

This awareness of our fundamental freedom separate from and progressively counter to that of the free market has thus opened up new avenues for socially conceiving and practically engaging with freedom. To this effect, it has produced a palatable optimism that the system can be changed and that we as humans can be the driver of this transformation. This optimism ranges from large-scale political movements on both the Left and the Right seeking a "revolution" for upending a once sacred status quo to small-scale efforts to reconfigure social relations away from those dictated by the market such as the sharing economy. Emerging, hence, in existential terms is a despairing capitalism, its hopeful destruction as essentialized way of viewing and acting in the world and the reawakening of our existential freedom to choose alternative modes of existence in its place.

Such radical optimism is itself confronted with widespread feelings of existential anger and abandonment associated with the loss of our previous inviolable faith in market salvation. There is a definite wrathful anger and anxiety that comes from realizing that our idols are false and beliefs hollow. Indeed, the acknowledgement that "God is Dead" is usually as mournful as it is gleeful. It is exactly in this mourning for a divine force that is slipping away in which bad faith grows and prospers. Many continue to believe that the free market can deliver us from our economic and social evils. If only we asked it for to forgive us for our heresy and recommitted to it with a renewed fervour, our sins could be atoned.

The perhaps most pressing and profound challenge of our times, then, is to break free from the free market, to dispose of its essentialized market freedom and be willing to seek out new ones emancipated from any pseudo-religious blinders. It is to take responsibility for the shared condition of our existence and commit to refashioning society according to different principles. It is a stark but also optimistic that the choice of history remains open and more importantly is ours still to make and remake. Significantly, it also points to the need to existentially acknowledge that we are not only condemned to be free but also to decide upon the type of freedom that we will temporarily live with.

The New Foundations of Freedom

The challenge of freedom, one that always lies dormant in wait, never obviously occurs in a vacuum. They may appear as moments or often sudden radical breaks, or they may be fueled by the general decline in the "truth" and promises of existent social order. The current existential desires to be free from the free market are no different. While its structural roots are certainly found in the unregulated greed of a financial system run amok, its seeds spring from an attempt to exist in a world seeking to transcend current realities for new ones. To this effect, post-structuralism has the potential to be a key theoretical perspective for understanding and guiding these novel attempts at freedom.

On the surface, the relationship between freedom and post-structuralism is if not completely inimical then at the very least not immediately apparently complementary. Indeed an oft-repeated critique of this admittedly broad theoretical perspective is its pronounced lack of robust conception or defence of any sort of inherent liberty. It is argued that

...the denial of an autonomous subject leads to the denial of any meaningful concept of freedom, which again leads to the impossibility of any emancipatory politics. When there is no authentic subjectivity to liberate, and power as the principle of constitution has no outside, the idea of freedom becomes meaningless. Since we are always products of codes and disciplines, the overthrow of constraints will not free us to become natural human beings. Henceall that we can do is produce new codes and disciplines. (Oksala 2009: 1–2; note that the author does not endorse this view but is only presenting it)

Its basis in social constructionism leaves it vulnerable to attacks of political relativism and a less than full-throttled belief in the sanctity of freedom of any kind. As renowned philosopher Todd May (1994: 6) observes (before morally defending post-structuralism on consequentialist grounds as having a theory of power that is both "creative and pervasive") "Critics of the post-structuralist approach to political philosophy, especially those associated with Critical Theorists ... have seen the lack of moral grounding for post-structuralist claims as one of the most problematic areas of its thought".

These critiques, while overly broad and commonly polemic, have nevertheless some credence. Putting aside the shared mistakes of many on the Left in confusing revolution with dictatorship, it is telling that so much of post-structuralism's main proponents normatively revert back into what may appear to be a relatively conventional account of liberal or republican forms of democratic freedom and tolerance. However, it also provides the philosophical tools for deepening our critical engagement with existential freedom. A central value of post-structuralism is that it frees us from the tyranny of inscriptive and essentialized structures and meanings. In place of inherent freedom, it offers a more emancipating relative freedom. Put differently, in its constant unsettling of truth and innate ideas of human nature, it affords us greater opportunities to embrace a conscious existence as what can be termed as "beings-for-ourselves".

Importantly, to assert that we are free from structures and meanings does not imply that we completely dispose of them. Conversely, it demands that we treat them as socially constructed—and therefore utterly changeable—frameworks that shape our existence. Post-structuralism, in this spirit, provides individuals and communities the critical perspective to begin undoing our perceived essences and uncovering existing gaps of freedom. While such gaps are always with us, post-structuralism permits us to encounter them existentially as opposed to essentially. We do not seek to overcome through any innate or intrinsic form of freedom. By contrast, we view them as an opportunity to discover new freedoms and fresh modes of existence.

Moreover, this is no mere abstract concern. To once again paraphrase Sartre, we are "condemned to contingency". No matter how much we seek to escape the mutability of the world, to naturalize and essentialize it, we will always ultimately be confronted with an existence that defies any complete predetermination or teleology. Yet this also proves true in the reverse, as that which is deemed necessity must continually contend with the forces of contingency, seeking always to explain its incompleteness and cover over constantly emerging existential gaps.

What post-structuralism does, in turn, is to reveal that the only true necessity is contingency and the existential freedom that this implies. If this realization can be overwhelming and dispiriting it can also be liberating and invigorating. It grants us novel opportunities to build alternative social foundations in the space of these gaps. Oliver Marchant speaks, hence, of a "foundational difference", referring to the persistent and paradoxically productive tension that exists between the inability of any structure or truth to be total or final and our just as certain inability to ever live completely free from them. It is from such dynamic roots, hence, that we are able to perpetually refashion our existence and forge ever newer foundations of freedom.

References

Alford, R., & Friedland, R. (2011). *Powers of Theory*. Cambridge, GBR: Cambridge University Press.

Althusser, L. (1969). *For Marx*. Verso.

Anievas, A., & Nişancıoğlu, K. (2015). *How the West Came to Rule*. London: Pluto Press.

Appleby, J., Hunt, L., Jacob, M., & Markoff, J. (1996). Telling the Truth About History. *Contemporary Sociology, 25*(1), 130.

Bernstein, H. (1971). Modernization Theory and the Sociological Study of Development. *The Journal of Development Studies, 7*(2), 141–160.

Bloom, P. (2016). *Authoritarian Capitalism in the Age of Globalization.* Edward Elgar Publishing.

Brown, W. (2017). *Undoing the Demos: Neoliberalism's Stealth Revolution.* Boston: MIT Press.

Butler, E. (2013). *Foundations of a Free Society.* London: Institute of Economic Affairs.

Crouch, C. (2012). There Is An Alternative to Neoliberalism that Still Understands the Markets. *The Guardian.*

Eaton, G. (2017). The Tories Are Stuck in No Man's Land on Austerity. *The New Statesman.*

Elliott, L. (2017). Theresa May to Champion Free Market in Bank of England Speech. *The Guardian.*

Elliott, L. (2017). Theresa May to Champion Free Market in Bank of England Speech. *The Guardian.*

Ghemawat, P. (2017, February). Even in a Digital World, Globalization Is Not Inevitable. *Harvard Business Review.*

Hall, T. (2011). The Triple Bottom Line: What Is It and How Does It Work? *Indiana Business Review, 86*(1), 4–8.

Harvey, D. (2005). *A Brief History of Neoliberalism.* Oxford: Oxford University Press.

Hayek, F. A. (2014). *The Road to Serfdom: Text and Documents: The Definitive Edition* (Vol. 2). New York: Routledge.

Kaletsky, A. (2017). The Crisis of Market Fundamentalism. *Social Europe.*

Kavanagh, D. (1987). *Thatcherism and British Politics: The End of Consensus.* Oxford: Oxford University Press.

Latham, M. E. (2010). *The Right Kind of Revolution: Modernization, Development, and US Foreign Policy from the Cold War to the Present.* Cornell University Press.

Leys, C. (1997). The British Labour Party Since 1989. In D. Sassoon (Ed.), *Looking Left: European Socialism After the Cold War* (pp. 17–43). London: I.B. Tauris.

Marx, K., & Engels, F. (2016). *The Manifesto of the Communist.* Narcissus.me.

May, T. (1994). *The Political Philosophy of Poststructuralist Anarchism.* Penn State Press.

Mishra, P. (2016). Welcome to the Age of Anger. *The Guardian.*

Mitchell, W., & Fazi, T. (2017). *Reclaiming the State.* London: Pluto Press.

Monasterios, E. (2018). Uncertain Modernities: Amerindian Epistemologies and the Reorienting of Culture. In S. Castro-Klaren (Ed.), *Blackwell Companions to Literature and Culture* (pp. 553–570). Malden, MA: Blackwell Publishing.

Oksala, J. (2009). *Foucault on Freedom*. Cambridge: Cambridge University Press.

Pocock, J. (2001). *Barbarism and Religion*. Cambridge: Cambridge University Press.

Sartre, J. (1948). *Existentialism and Humanism*. New Haven: Yale University Press.

Sherman, R. (2017). *Uneasy Street: The Anxieties of Affluence*. Princeton, NJ: Princeton University Press.

Temperley, H. (1977). Capitalism, Slavery and Ideology. *Past and Present, 75*(1), 94–118. https://doi.org/10.1093/past/75.1.94.

Tipps, D. (1973). Modernization Theory and the Comparative Study of National Societies: A Critical Perspective. *Comparative Studies in Society and History, 15*(2), 199.

Venugopal, R. (2015). Neoliberalism as Concept. *Economy and Society, 44*(2), 165–187.

3

Capitalism's Existential Crisis: Producing Existential Freedom

In 2008, the world almost came crashing down. The once sacred belief in the power of finance was close to being overturned in a matter of months. Major banking institutions fell overnight and the certainty of financial progress seem to have reached a dead end. And sunny in a sweet historical irony, it was the hated big government that was suddenly expected to come to the rescue. More than just an economic downturn it is a complete crisis of confidence in the financial system itself.

A decade following the crises much has changed and much as stayed the same. Far from giving birth to a total economic transformation, the financial sector has perhaps remained as powerful and dominant as ever. Predictably inequality continues to be chronic and economic insecurity rampant. However, there are clear signs of change on the horizon. Anti-establishment fuelled populism is upending politics globally. Established democracies appear under threat by virulent far-right nationalism. Ideologically, socialism has sparked to life again from the ash heap of history. The victory of capitalism in the free market was no longer obvious or secure.

An increasing number of people were looking for radical economic alternatives. Socialism stood as an attractive option to a generation of

© The Author(s) 2018   41
P. Bloom, *The Bad Faith in the Free Market*,
https://doi.org/10.1007/978-3-319-76502-0_3

voters plagued by a lifetime of chronic unemployment, entrenched elitism, an ineffective public sector, and an uncertain future. "Socialism is back", reported social commentator John Quiggin (2017: N.P.),

> much to the chagrin of those who declared it dead and buried at the "end of history" in the 1990s. When the New Republic, long the house organ of American neoliberalism, runs an article on The Socialism America Needs Now, it's clear that something has fundamentally changed. The soft neoliberalism represented by Tony Blair, Bill Clinton and Paul Keating has exhausted its appeal, and not just in the English-speaking world. Throughout Europe, new movements of the left have emerged to challenge or displace social democratic parties discredited by the austerity politics of the last decade.

Reflected was a profound existential crisis to capitalism. It represented much more than a simple ideological struggle or a temporary economic downturn. Indeed

> The really bad news is that there is no structural or fundamental change to the fourth epoch global capitalism on the horizon that will be positioned to change the situation and become its success or replacement. It is an existential crisis. (Savall et al. 2017: xiv)

It was a serious question of whether humans can still save their collective destiny, where they stuck with a present that would stretch onward into an exploitive and unchanging forever. Was the current system permanent regardless of its efficacy or continued desirability? The objective laws supposedly underpinning the free market were progressively shown to be false.

At stake, therefore, was more than simply an economic challenge. Knowledge and meaning themselves were up for grabs. The previously unquestionable had turned in front of the world's eyes to be built upon very shaky truths indeed. According to economist Christian Arnsperger (2010),

> When I talk about an existential crisis, I mean that, in fact, the roots of this crisis are existential and found in each one of us. So we could talk about an anthropological crisis … one cannot do without the economy, but one can

and one will have to do without capitalism. This existential crisis of the economy is a truly essential crisis of capitalism, the symptom of a profound malaise. The existential crisis of the economy we are participating in today rests primarily on a crisis of confidence. People consume less, have a tendency to slow down accumulation and investment. But what stands out from my research work in philosophy of the economy is that capitalist consumption, investment and accumulation are themselves a symptom of the lack of fundamental confidence in life and in the future.

What has thus emerged following the crash was a decade-long loss of faith in the sacred tenants of financial capitalism and the free market.

Nevertheless it also opened up new social possibilities. More precisely it created the very conditions for humanities existential freedom to once again arose from its dogmatic slumber. This chapter will explore the existential crisis following the 2008 financial crash—specifically the feelings of alienation that rose up in its wake and the resurgence of anti-capitalist perspectives such as Marxism. It will use this analysis to critically interrogate what freedom is actually produced by the free market and how this reflects a broader dialectic of existential freedom historically.

Capitalism in Crisis

The financial crisis was a potential paradigm changer. It threatened the entire status quo and its capitalist religion of the free market and financialization. Quoting renowned progressive financier George Soros (2008: N.P.) in none other than the *Financial Times*

> We need new thinking, not a reshuffling of regulatory agencies … For the past 25 years or so the financial authorities and institutions they regulate have been guided by market fundamentalism: the belief that markets tend towards equilibrium and that deviations from it occur in a random manner…. Regulators ought to have known better because it was their intervention that prevented the financial system from unravelling on several occasions. Their success has reinforced the misconception that markets are self-correcting.

In the intervening years such change has thus far largely failed to materialize. Rather there has been a renewed embrace of the free market and the continued necessity of capitalism as an economic system. This is the age of austerity not revolution.

Much of this reaction can be explained by the strategic manoeuvring of elites to popularly legitimize the entrenched order that they literally and figuratively profit from. Critically and persuasively economist Phillip Mirowski reveals exactly this in his aptly titled book *Never Let a Serious Crises Go to Waste*. He speaks almost apocalyptically, in this respect, asking us to

> conjure, if you will, a primal sequence encountered in B-Grade horror films where the celluloid protagonist suffers a terrifying encounter with doom yet on the cusp of disaster abruptly awakes to a different world which initially seems normal but eventually revealed to be a second nightmare more ghastly than the first. Something like that has become manifest in real life since the onset of the crises which started in 2007. (Mirowski 2013: 1)

He then critically and with prescient insight showed how the crisis has failed to produce any serious fundamental economic rethinking or transformation.

While profoundly valuable this analysis perhaps too readily overlooks the deep structural and affective attachment individuals have to this exploitive system. The free market for all its obvious flaws still remained to many the only alternative available. Its fall from grace represented a deep fear that their very material prosperity and even survival were now at risk. This reflects a crucial paradox of capitalism first gestured to by Marx. A continual issue confronting those who would like to challenge or even wholesale change a market economy is why people continue to invest their time, money, and energy into its preservation. If the entire system is in the short-term exploitive and in the long-term ruinous for everyone, why remain so attached to its reproduction?

While the answer to these questions exceeds the scope of any one single analysis, a key factor is that capitalism is deemed central to people's well-being and sense of freedom. The may not like working or their working conditions but it's far better than the alternative of being hungry, homeless, or unemployed. Moreover, it is truly hard to imagine alternative ways besides those offered by the free market of actively improving our condition and to an extent controlling our personal historical destiny.

If there is such a thing as "false consciousness" as suggested by Marxism surely it is found in this transformation of this material system into a perceived social and psychic need. Using other language, one's entire sense of self is wrapped up in their job, career, and personal opportunities to provide. Without the benefit of the market, there simply is no chance for the society to function or for the people to feel empowered.

The financial crisis put into jeopardy both this sense of socio-economic security and the prevailing sense of agency that accompanied it. Cast in this light, it is understandable why the immediate post-crisis discourse of recovery was and to some extent remains so appealing. In this context what was being salvaged was not so much a broken system but a guaranteed future of individual and collective prosperity now placed at risk.

The economic insecurity so pervasive in this era, hence, is so much more comprehensive than the material lack of food or shelter (though these should not be overlooked). It is a profound insecurity in respect to one's identity and place in the world. Reality, quite literally at times, stops making sense without the market to sustain it. As such shared narrative of progress become devalued so do the personal journeys of those who populate it.

At stake is a deeper existential anxiety—a confrontation with the world devoid of any inherent meaning or order. The influential twentieth-century German philosopher Martin Heidegger (1996) describes the anxiety as an encroaching feeling of "foreignness", of no longer being at home in the world. It is to suddenly be a stranger in a strange country. Whereas previously the long hours, the chronic fear of losing your job, the thrill of getting a new one, the inevitable disappointment after this initial jubilance, the banal, everyday realities of organizational injustice—these were familiar and comfortable even in their exploitation. Now we wake up to a world that no longer seems so sturdy or so comfortable. In the words of modern critical philosopher Simon Critchley (2009: N.P.),

What is first glimpsed in anxiety is the authentic self. As the world slips away, we obtrude. I like to think about this in maritime terms. Inauthentic life in the world is completely bound up with things and other people in a kind of "groundless floating"—the phrase is Heidegger's. Everyday life in the world is like being immersed in the sea and drowned by the world's suffocating banality. Anxiety is the experience of the tide going out, the seawater draining away, revealing a self stranded on the strand, as it were. Anxiety is that basic mood when the self first distinguishes itself from the world and becomes self-aware.

It is utterly and tragically predictable then that in the wake of this crisis, in the midst of such pervasive economic and social insecurity, there would be fears that the "old ways" were slipping away. It is indeed reminiscent of existential anguish. It is a terrifying recognition that reality can change overnight, meaning can become meaningless even more quickly, and that just as certainly we have little or no chance to prevent these once considered sacred truths from disappearing. The pull of the past for an idealized time when things made sense and everything was in its proper place is strong.

The ultimate reward of the free market—its most true and real utility is found not in its promise of material wealth beyond our wildest imagination. Nor is it linked exclusively to the freedom it supposedly provides. Rather, it is found in the stability it offers as it keeps our existential fears at bay. We accept, in this respect, a pervasive economic insecurity for a precarious existential security. Yet it is precisely such existentially fraught times that are ripe for revolution.

The Search for a Revolutionary Method: Existentialism and Marxism

At the heart of capitalism lies a profound tension—an irony of historic proportions. It asks us to exchange our existential freedom for its more limited market freedom. The reward for such a transaction is that it holds at bay existential anguish, the swindle is that it can only do so for long. It is precisely for this reason why Marxism for all its own historic failings remains so timely and timeless, timely, in that it provides an alternative to a capitalist system in crises, timeless, in that it still holds the promise of existential liberation.

In the opening chapter of his seminal later work "The Search for a Method", Sartre directly grapples with the fraught relationship between existentialism and freedom. To do so, he sets out an ambitious theoretical course, putting forward an entire conceptual edifice for understanding philosophy, ideology, and history. While obviously a full treatment of this analysis lies far beyond the scope of this book, its main points raise pertinent clues about the fundamental importance of existential freedom and its potential radical contemporary implications. In particular, it sheds light on the tension filled but productive relation of existentialism to Marxism.

Sartre begins by reapproaching what is meant by the concept of philosophy itself. Rather than searching for any singular definition of this term, he argues instead that there are philosophies. Crucially an active philosophy unites existing knowledge together, providing a totalistic framework for viewing and acting within the world. Accordingly,

The philosopher effects the unification of everything that is known, following certain guiding schemata which express the attitudes and techniques of the rising class regarding its own period and the world ... These achievements of knowing, after having been first bound together by principles, will in turn-crushed and almost undecipherable-bind together the principles. Reduced to its simplest expression, the philosophical object will remain in "the objective mind" in the form of a regulative Idea, pointing to an infinite task. (Sartre 1963: 4)

Concretely, this all-encompassing basis for thought and action represents the emergence of a "rising class" becoming "conscious of itself". He notes that

A philosophy is first of all a particular way in which the arising class becomes conscious of itself. This consciousness may be clear or confused, indirect or direct. At the time of the *noblesse de robe* and of mercantile capitalism, a bourgeoisie of lawyers, merchants, and bankers gained a certain self-awareness through Cartesianism; a century and half later, in the primitive stage of industrialisation, a bourgeoisie of manufacturers, engineers, and scientists dimly discovered itself in the image of universal man which Kantianism offered to it. (Ibid.: 4)

As the above quote shows he charts, in this regard, a modern history of philosophy—starting with the mercantilism underpinning the ideas of Locke and Rousseau, followed by the industrialization informing Kant and Hegel, and culminating in the twentieth century composed of Marxism and the proletariat.

Sartre's perspective, of course, suffers from a slight case of economic reductionism. A philosophy, in this reading, reflects its economic conditions, granting it a supposed exhaustive social reality. Nevertheless, it gestured to the broader social significance of existentialism generally. He declares,

> Thus a philosophy remains efficacious so long as the praxis which has engendered it, which supports it, and which is clarified by it, is still alive. But it is transformed, it loses its uniqueness, it is stripped of its original, dated content to the extent that it gradually impregnates the masses so as to become in and through them a collective instrument of emancipation. (Ibid.: 5)

Notably, in the ways that present-day capitalism has produced its own neoliberal philosophy populated by a growing class of "entrepreneurial consumers".

Sartre, for this reason, critiques existentialism for never achieving the status of a totalistic social philosophy. Instead it has always remained in his view merely as a form of ideological critiques against such philosophical systems. In all of its previous permutations it existed only in opposition to a more dominant worldview. It also suffered from at points being to inwardly focused, localizing its insights at the level of the individual rather than the mass of society. However, it is precisely as an ideology though that existentialism most critically intervenes and transforms Marxism.

> Thus the autonomy of existential studies results necessarily from the negative qualities of Marxists (and not from Marxism itself). So long as the doctrine does not recognize its anemia, so long as it founds its Knowledge upon a dogmatic metaphysics (a dialectic of Nature) instead of seeking its support in the comprehension of the living man, so long as it rejects as irrational those ideologies which wish, as Marx did, to separate being from Knowledge and, in anthropology, to found the knowing of man on human existence, existentialism will follow its own path of study. (Ibid.: 181)

While Marxism had begun as "moment" of freedom in its initial European revolution of 1848 and the Bolshevik revolution in 1917, it had decayed by the second half of the twentieth century into a rigid and dogmatic discourse stagnating politics, culture, and even economics. Sartre refers to this philosophical degradation as a type of "idealization" since it demands that every action, historical event, and social idea conforms to its strict and narrow view of reality. It is from this embrace of orthodoxy that the once vibrant philosophies fall prey to and ultimately become the source of repression.

To contemporary readers, this critique of Marxism sounds obviously all too familiar. There is a tendency, perhaps understandably, for those on the Left to resist this historical judgement, to try to historically salvage really existing socialism from its harsh truth of gulags, man-made famines, and autocratic oppression. However, there is an equal and arguably more ominous danger of exploiting these valid condemnations to champion an equally idealized and destructive modern free market. What Sartre is offering, therefore, is a deeper critical questioning of how much existential freedom there is in any prevailing philosophical system at any given time of its social dominance.

How much existential freedom, therefore, remains left in the free market, if there was any substantially to begin with? More precisely, to what extent does it allow for humans to challenge, reinterpret, change, and transform their existent social conditions? What agency does it provide individuals and communities to question their existence and radically choose new ways of life? Or is the free market merely another repressive status quo who demands that all things fit within its idealized worldview?

Producing Freedom

Marxism almost as a rule has been rather allergic to liberal definitions of intrinsic freedom. For the true believers, it is scoffed at as mere bourgeoisie facades, myths to keep the masses in line and receptive to their own economic exploitation. Indeed, when Marxist freedom is spoken of it is usually referred to in rather oblique terms—as the emancipation from capitalism and despite its non-utopian pretensions as a secular fantasy of a coming Communist society. In practice it's most vibrant freedom has been found in its possibilities of producing revolution. It is tempting, therefore, to dismiss Marxism on the grounds of freedom alone—and there are certainly historical reasons for doing so. It is perhaps just as tempting to embrace it as the vehicle upon which we still have the freedom to bring about whole-scale revolutionary change to our collective existence. The vaunted third way (see Giddens 1993) is to dream about possibilities of constructively combining the valuable elements of Marxist and capitalism. There is a fourth way, however, which is to understand what freedom capitalism materially manufactures as well as the existential gap it produces.

Central to Marxist theory is the concept of production. While Marx put forward a still valuable labour theory of value, it is fundamentally contained and operating within a historically specific mode of production. He writes that

> The way in which men produce their means of subsistence depends first of all on the nature of the actual means of subsistence they find in existence and have to reproduce. This mode of production must not be considered simply as being the production of the physical existence of the individuals. Rather it is a definite form of activity of these individuals, a definite form of expressing their life, a definite mode of life on their part. As individuals express their life, so they are. What they are, therefore, coincides with their production, both with what they produce and with how they produce. The nature of individuals depends on the material conditions determining their production. (Marx 1970: 42)

Marxism, hence, provides posterity with a materialist conception of history based on class struggle. Here economic production is transformed, at least abstractly, into a form of social reproduction and historical overdetermination. Capitalism will invariably create competing classes, pitting the capitalist and the bourgeoisie on the one side and the proletariat and to a lesser extent the peasantry and the underclasses (the so-called lumpenproletariat) on the other. This struggle reaches its inevitable conclusion as the profit drive lead to ever higher unemployment and finally working class revolt followed by the triumph of capitalism over communism. Consequently, Marx declares in the Preface to his first German edition of his grand work *Capital* (1999):

...here individuals are dealt with only in so far as they are the personifications of economic categories, embodiments of particular class-relations and class-interests. My standpoint, from which the evolution of the economic formation of society is viewed as a process of natural history, can less than any other make the individual responsible for relations whose creature he socially remains, however much he may subjectively raise himself above them.

Marxism certainly would seem then like a strange place from which to derive a more comprehensive philosophical framework for conceiving and enacting existential freedom. His materialist account leaves little room it would appear for either individual or historical agency. Further, in direct contrast to Sartre, he views consciousness as an outgrowth of its materialist conditions, as the quote above makes abundantly clear. Nevertheless, contained within it is a potential opportunity for reconsidering freedom—one that at once accepts its historical giveness and also highlights its expansive possibilities. In constructing his theory, Marx tellingly posits the social creation of classes and the opportunities these deterministic economic changes have for enlarging the scope of existential freedom. He argues in his unfinished third (and last) volume of *Capital* (2007) that

> Freedom in this field can only consist in socialised man, the associated producers, rationally regulating their interchange with Nature, bringing it under their common control, instead of being ruled by it as by the blind forces of Nature; and achieving this with the least expenditure of energy and under conditions most favourable to, and worthy of, their human nature. But it nonetheless still remains a realm of necessity. Beyond it begins that development of human energy which is an end in itself, the true realm of freedom, which, however, can blossom forth only with this realm of necessity as its basis.

Namely, the rise of the proletariat redefined how freedom was dominantly framed, focusing it on the increase in labour power. By emphasizing the significance of ownership, the power to define and manage organizational relations, and the need to make profit, he offered a new lens for seeing what freedom currently was and what it could eventually become. "There is in every social formation a particular branch of production",

Marx (2005: 146) proclaims, "which determines the position and importance of all the others, and the relations obtaining in this branch accordingly determine the relations of all other branches as well. It is as though light of a particular hue were cast upon everything, tingeing all other colours and modifying their specific features."

Thus while Marxism is often linked, and rightfully so, to themes of power, it also speaks to the fresh potential capitalism held for reconfiguring and expanding individual and collective agency. Capitalism as with any dominant social system produces a range of specific existential gaps for humans to fill and eventually overcome. He declares, in this regard, that

> No social order is ever destroyed before all the productive forces for which it is sufficient have been developed, and new superior relations of production never replace older ones before the material conditions for their existence have matured within the framework of the old society. Mankind thus inevitably sets itself only such tasks as it is able to solve, since closer examination will always show that the problem itself arises only when the material conditions for its solution are already present or at least in the course of formation. (Marx 2004)

Free market societies are marked by three in particular—the first resides at the individual level found in the ability to personally change one's circumstances through hard work, tale, and taking advantage of market opportunities. The second is found at the level of organizations, emphasizing the capacity of people to be brought together, incentivized, and managed in such a way to increase economic growth and capitalist innovation. The third, but by no means the least in terms of significance, is the collective agency to alter our shared social conditions, it is always worth remembering that capitalism was born in the ferment of revolution and grew up believing in its own historical task of transforming the world.

The Existential Dialectic of Freedom

Freedom is inexorably linked to economic and social production. The base-superstructure model often attributed to Marx whereby the economy structurally determines the social and political is obviously to crude a formulation to understand the rich and complex contemporary world. Nonetheless, it holds revealing clues about power relations shaping freedom. To this end, the existentialism drives history forward and is as yet also always ultimately contained and defined by it. What is produced, thus, is an ongoing dialectic of freedom.

A central tenet of existentialism is that existence precedes essence, as discussed in the previous chapter. Crucial to this insight is then that freedom necessarily is absolutely formative to essence. Who one is—their very self—is constituted by a fundamental freedom. This basis of freedom is two-fold at once universal and relative to a given time and place. The universal is the shared desires of humans to have some power over their actions and historical destiny—whether individually, collectively, or both. The relative aspect is the historically specific instantiations of this longing—made possible and manifest by a dominant social order. Thus, for example, in considering the growth of democracy, he argues "All forms of the state have democracy for their truth, and for that reason are false to the extent that they are not democracy" (1970: 31).

Thus who one is and who one seeks to be is determined, though never completely so, by the world in which they inhabit. As Marx rightly points out, material reproduction and labour form a key part of this determination. Yet what it also reflects upon is their production of diverse modes of freedom through which this essential self is given form and propelled forward into an active existence. This diversity is perhaps most highlighted within contemporary thought by the ideas of "social reproduction theory". Drawing on a range of critical perspective, it views material reproduction as the creative basis for a wide range of social identifications and "modes of being" to emerge. Quoting one of its leading thinkers Tithi Bhattacharya (2017: 19)

(social reproduction theory) reveals the essence category of capitalism, its animating force, to be human labour and not commodities. In doing so, it exposes to critical scrutiny the superficiality of what we commonly understand to be "economic" processes and restores to the economic process its messy, sensuous, gendred, race, and unruly component: living human beings, capable of following orders as well as flouting them.

There is therefore a profound dialectic at play between existence and essence, of which existential freedom is always at its centre. In this respect, individuals and events are to use a term first populated by the French Marxist Althusser, that is, "overdetermined"—their origins and development traced back to the social mode of freedom from which they have derived. In particular, it refers to the persistence and relative autonomy of existing "super-structure" forces such as cultural norms and political institutions for impacting, and to an extent possibly deforming, the potential for revolutionary transformation. Consequently,

> the overdetermination of any contradiction and of any constitutive element of a society, which means: (1) that a revolution in the structure does not ipso facto modify the existing superstructures and particularly the ideologies at one blow (as it would if the economic was the sole determinant factor), for they have sufficient of their own consistency to survive beyond their immediate life context, even to recreate, to 'secrete' substitute conditions of existence temporarily; (2) that the new society produced by the Revolution may itself ensure the survival, that is, the reactivation of older elements through both the forms of its new superstructures and specific (national and international) 'circumstances'. (Althusser 1969: 115–116)

As such, it is the contradictions themselves, between what exists and concretely the emergence of what could exist, that drive forward history (even if it is always, according to Althusser "in the last instance" reflective of its economic conditions). He observes,

> The specific difference of Marxist contradiction is its unevenness, which reflects its conditions of existence, that is the specific structures of unevenness (in dominance) of the ever pre-given whole which is its existence. Thus understood, contradiction is the motor of all development. Displacement and condensation, with their basis in its overdetermination, explain by their dominance the phases (non-antagonistic, antagonistic, and explosive) which constitute the complex process, that is, "of the development of things". (Ibid.: 217)

Consequently, in the case of the free market, its freedom is to be employable; then existence is directed at gaining new or more skills that at once can be an empowered and free employee. Marx, in this regard, writes of the

...insipidity of the view that free competition is the ultimate development of human freedom; and that the negation of free competition = negation of individual competition and of social production founded on individual freedom. It is nothing more than free development on a limited basis—the basis of the rule of capital. This kind of individual freedom is therefore at the same the most complete suspension of all individual freedom, and the most complete subjugation of all individuality under social conditions which assume the form of objective powers, even of overpowering objects— of things independent of the relations among individuals themselves. (Marx 2005: vi)

Likewise, if democracy is the cornerstone of achieving broad-scale change, then political existence is found in the ideological and electoral struggles of winning sovereign power.

At stake is how freedom is confined to and channelled within overdetermined directions—so that the absolute freedom of action is bound to the specific options one has available to them as well as the perceived dominant means provided for pursuing them. Further, in the constant pursuit of this freedom, the system which it is based comes to be reproduced and even strengthened. Existence, in this regard, becomes ever more crystallized into a perceived inherent essence. However, freedom is never static, and the yet untold possibilities of existence never cease to assert themselves. The success of a status quo is its capacity to maintain its promise that the potential for human life is still limitless within its philosophical horizons. All the while forcing them to conform and restrict their decisions to its narrow limits of Being. As such the free market promises that people can do and achieve anything just so long as it is profitable and fiscally responsible.

There is therefore an always present productive tension—whereby the desire for freedom must contend with the social restrictions placed upon it. Crises occur not only when a market crashes or the economic contradictions of capitalism become more obvious and pressing but also when the longing for existential freedom exceeds the limits of the current system en masse. The Open Marxist describes the development of capitalism, its rise and fall, as a matter of crises management, emphasizing the ability of elites to co-opt popular anger and demands for change to conform to their ideas and practices of a market economy (see especially Bonefeld et al. 1992; Burnham 1994). It reflects, in this respect, the wider array of "relations of production" including the creation of a "radicalized working-class subject" (Bell and Cleaver 1982).

Critical here is the socialized and analogous ways that a crisis becomes an opportunity for those in power to assert the continued potential for experiencing freedom linked to the status quo. Claims for reforms are a clear signal not only that a given order can be improved but that the possibilities for it be changed and for humans to reassert their control over their historic fate has not been extinguished. Importantly

> such a political reading of crisis theory eschews reading Marx as philosophy, political economy or simply as a critique. It insists on reading it from a working-class perspective and as a strategic weapon within the class struggle. (Bell and Cleaver 1982: 191)

It is perhaps tempting to think that this occurs completely at the whims of history. That it follows no set pattern—that the course of freedom flowed with no clear rhyme or reason. Yet while by no means historically predetermined it does adhere to a certain regularity. More precisely, it is possible to identify a certain general cycle of freedom, one that echoes the movement from philosophy to idealization introduced earlier by Sartre. What begins as a philosophy—a novel complex system of thought and action for experiencing freedom—becomes over time a sedimented and dogmatic status quo inhibiting new forms of freedom from emerging. Hence, it is in the evolution from existence to essence, possibility to orthodoxy, that this fundamental freedom is diminished and born anew.

This cycle extends far beyond abstract debates over liberty. It reflects an updated version of the materialist dialectic proposed by Marx and expounded by his proponents. As new material practices emerge, they reveal the current concrete limits of existing freedom and the possibilities for their genuine negation and overcoming. In doing so, it reawakens humanity's existential potential and points the way for new modes of freedom to become real and spread. Marx hints at just such a dynamic relation of dialectical freedom in his discussion of working-class organization linked to their overall politicization and emergent revolutionary "class consciousness":

> When communist *artisans* associate with one another, theory, propaganda, etc., is their first end. But at the same time, as a result of this association, they acquire a new need—the need for society—and what appears as a means becomes an end. …the brotherhood of man is no mere phrase with them, but a fact of life, and the nobility of man shines upon us from their work-hardened bodies.

These instances, thus, are often at first small but can with time and under the right circumstances come to loom large.

Present is an existential dialectic of freedom. In moments where prevailing freedoms become opposed by new ones, what was once not even dreamed of is transformed into a revolutionary aspiration and then a manifest certainty. Indeed, freedom at its perhaps most fundamental can be understood as the emancipation from a tyrannical essence for the liberty to rediscover the possibilities of existence.

Capitalism's Existential Crisis

It is widely acknowledged that in the past half-decade the previously unthinkable has occurred. Capitalism is deeply in crisis. Young people in particular in once considered free market strongholds such as the UK and US are turning away from the system and advocating for progressive radical changes. Spotlighted quite naturally are material issues of rising inequality and economic insecurity. These are further exacerbated by the long-standing problems of racism, sexism, xenophobia, and homophobia. However, buried deep but burning bright is a resurgent demand for existential freedom in an age when the free marketed seems suddenly the furthest thing from freedom. Reflected is the idealization of capitalism, its transformation into a regulative and restrictive social "essence".

Beginning in the late twentieth century there has been (at least for the privileged global north) a veritable "new spirit of capitalism" (Boltanski and Chiapello 2005). This rather bold term indicates the shift in the emphasis of capitalist production away from the overly regulated and towards the promotion and harnessing of individual creativity. Represented is a fresh "ideology that justifies engagement in capitalism" (Ibid.: 8). This may seem at odds with the crass materialism of neoliberalism beginning in the 1980s—where wealth was valued for its own sake. However, it crucially channelled this human creativity towards accruing profit and capitalist innovation.

Implicit in this new vision was the creative impulse to go beyond neoliberalism, to experience more than what the free market had to initially offer. It represented the attempted resolution of a crucial capitalist contradiction—a system that is sustained by a rhetoric of freedom and materially by strict regimes of conformity and regulation. Indeed it is a contradiction that has expanded into every corner of our social existence. While neoliberalism is often accused of economizing society, politics, and even our personal life, this creative aspect is too often forgotten. It is about how to innovate one's very existence using fiscal norms and techniques. Ong (2007: 3) thus refers to neoliberalism as a "mobile technology" observing

> the very conditions associated with the neoliberal—extreme dynamism, mobility of practice, responsiveness to contingencies and strategic entanglements with politics—require a nuanced approach, not the blunt instrument of broad categories and predetermined elements and outcomes … Neoliberalism is conceptualized not as a fixed set of attributes with predetermined outcomes, but as a logic of governing that migrates and is selectively taken up in diverse political contexts.

Importantly, this extends to the achievement of traditionally non-market ends—such as personal well-being and social justice. Capitalism is no longer simply interested in profit. Rather it is an expansive system for creatively achieving our dreams, the means to these diverse ends. It offers the promise of achieving "a balanced life" and being able to work "anytime, anywhere". It promises autonomy, flexibility, and the right to choose how one spends their times. Thus within its own limited ideology, the free market projects an image of itself as limitless in relation to human possibility.

Crucial then to the contemporary free market was a discourse of individual empowerment. It was meant to give people the knowledge, tools, and inspiration to be independent and freely make their own decisions about their existence. Of course such empowerment is easily critiqued as a tool of elites, one whose ultimate purpose is the increasing of efficiency and greater work intensification. It represented a veritable "wellness syndrome" reflecting a neoliberal self

> that is often best equipped to meet the contradictory demands of present-day capitalism: to be simultaneously extroverted and introspective, flexible and focused, adaptable and idiosyncratic. In other words, coaching does not just seek to improve people's wellbeing, or to teach them how to enjoy more. It is a technique aimed at reshaping the self. (Cederström and Spicer 2015: 15)

However, it also gestured towards the enlargement of market freedom, just as significantly, it signalled a profound reversal where people were meant to exploit capitalism and not vice versa. In other words, the goal was not more economic profits (at least in principle) but the use of these profits and economic techniques to serve the existential desires of capitalist subjects.

At stake was the reclassification of freedom, the production of the same system but as supposedly populated by new subjects. The traditional divide between ownership and labour—employer and employees—was gradually shifting within the public consciousness to a divide between empowered elites and the disempowered masses. The suddenly vilified 1% were castigated not only for their obscene wealth and perceived the unseemly use of it but also about their unfair power imbalances. These plutocrats and their elite political supporters could influence policy and pursue their happiness in a way that was simply unimaginable to the average person.

The populist upsurge wracking established democracies across the world is an attempt to countermand this dramatic difference in freedom. In this respect, class struggle is waged around a profound existential gap between social groups. Indeed, these social groups themselves are constituted by the existential freedom they wield and their ability to use existent freedom to their advantage. It was this class conflict that is leading capitalism's potential demise. Tellingly, elites have consistently referred to this system as a "sick patient" that must be "cured". Yet there appears to be a growing mass movement committed to euthanizing this patient—gradually and if necessary rapidly letting it out of its misery.

Witnessed, in turn, is the pronounced "Death of Homo-Economicus" as recently foretold by theorist Peter Fleming (2017). It was a demise signified by the desire for an existence that transcended mere profit-making and career advancement—and the anger that these non-market aspirations were the reserve of a privileged elite few. The still appealing dreams of "social entrepreneurship" or "smart solutions" are remnants of this growing battle for existential freedom. It is the latest tactic of a capitalism quickly running out of answers to assure a restless population that its possibilities are still infinite. What is being made starkly clearer though is that such limitless is simply unavailable in the world as brought to us by the free market.

The potential organic crisis of capitalism following the global financial crisis has deepened into what is fast becoming a full-blown existential crisis. The free market was the idealization of contemporary capitalism—its attempt to claim that it was a natural, objective, and permanent feature of human existence. However, its objective laws have been challenged as merely elite-constructed myths. Further there is a desire to break free from this seemingly inescapable neoliberal reality. There is a growing recognition that once sacred market truths are now up for debate and profound reconsideration. These gesture towards a more revolutionary desire to trade in the free market for the purchase of a more radical existential freedom.

The New Struggle for Existential Freedom

Capitalism is now facing one of its greatest historical threats. Its supposed economic laws, once revered in a spirit akin to physics, are now seen by an increasing number of people as closer to being a modern-day form sorcery. The cult of the free market is being exposed, creating an exciting but also fearful future. The resurgent popularity of socialism and to an extent Marxism speaks to this uncertainty. It also reflects renewed possibilities to collectively influence our shared material and social existence. Contained within this crisis are the incipient seeds of novel freedoms waiting to be born. Their sprouting represents signs of a capitalist spring giving life to a new existential struggle for freedom.

Indeed it is important to note that a funny thing happened on the way to a permanent capitalist world now and forever—people began discovering and experimenting with fresh types of social agency. Digital technologies, for instance, were paving the way for cutting-edge organizations and practices. While "platform capitalism" (Srnicek 2016) has been rightly criticized for contributing to an exploitive "gig economy", it is also linked to the potential for creating alternative and less market-based economic relations. The sharing economy is a prime example of these arising hi-tech possibilities. More broadly, these technologies have been a catalyst for a fundamental economic and social rethinking of sacred capitalist knowledge. A 2017 British Labour Party report hence calls for new hi-tech forms of "alternative ownership", proclaiming

> The economic system in Britain, in its current guise, has a number of fundamental structural flaws that undermine economic strength and societal well-being. The predominance of private property ownership has led to a lack of long-term investment and declining rates of productivity, undermined democracy, left regions of the country economically forgotten, and contributed to increasing levels inequality and financial insecurity. Alternative forms of ownership can fundamentally address these problems. These issues are all the more pronounced given the increasing levels of automation in our economy. Automation has an emancipatory potential for the country's population, but the liberating possibilities of automation can only be realized—and the threats of increased unemployment and domination of capital over labour only countered—through new models of collective ownership that ensure that the prospective benefits of automation are widely shared and democratically governed.

These new capabilities have contributed to a remaking of existential freedom en masse. The insurgent "radical leftist" movements across the US and Europe are a testimony to this long-dormant desire. While figures such as Corbyn and Sanders may ideologically be quite tame when compared to their supposed revolutionary forefathers, to a large extent their significant is to be found elsewhere in their symbolizing that radical social change is still possible and urgently necessary. It is the opening up of possibility to a brave new future free from capitalist labour. Quoting a recent editorial in *The Guardian* entitled "Post-Work: The Radical Idea of a World without Jobs", its author Andy Beckett (2018: N.P.) writes "Work has ruled our lives for centuries, and it does so today more than ever. But a new generation of thinkers insists there is an alternative."

Interestingly, these advances were tied to a quite profound call for reconsidering established market versions of freedom. Their broadly social democratic agendas reflect the growing recognition that freedom is contingent upon social security. Tellingly, the official linking of these values in the context of the War on Terror evolved into a profound economic critique against the basic tenant of the free market. Just as freedom needed to be secured against terrorism—safety being a fundamental condition of the spread democracy and liberty, at least in rhetoric—so it is understood that precarity and material insecurity inhibit the realization of free existence universally. This has catalysed calls, for instance, for a "universal basic income"—a policy practically unthinkable even a decade ago that is now being tried in a growing number of cities and even countries. While by no means a new idea in and of itself, it points to the articulation of the condition of possibility for the realization freedom more widely. More than simply representing the promotion of a different expression of freedom, it is a claim that to exist in all its rich specificity these underlying universal conditions must first be met.

Significantly, Sartre ultimately reached a similar conclusion, noting that perhaps the most defining feature of Marxism to revolutionary liberation is that for freedom to be fully realized freedom from scarcity is absolutely necessary. He declares

> As soon as there will exist for everyone a margin of real freedom beyond the production of life, Marxism will have lived out its span; a philosophy of freedom/will take its place. But we have no means, no intellectual instrument, no concrete experience which allows us to conceive of this freedom or of this philosophy. (Sartre 1963: 34)

What Marxism thus produces is a fresh understanding of the conditions of freedom as such. More precisely, it uncovers an existential dialectic of freedom which creates the social-material condition for this fundamental condition to be attained. Thus the transformation away from feudalism was in part a struggle for liberty against arbitrary authority, which revealed consent and self-determination as a universal prerequisite for existential freedom.

Similarly, the contradictions of capitalism exacerbated by the free market show clearly that scarcity makes true freedom impossible. And just as with feudalism and consent, this recognition is fatal to the very basis of capitalism itself and its association of liberty with a competition over resources, wealth, and social status. In this respect, it is a fundamental contradiction in that it reveals that the present and its essentialized form of market freedom are in fact fundamentally incompatible with our existential potential and the possibilities of new more dynamic ideas and practices coming into being.

References

Althusser, L. (1969). *For Marx*. London: Verso.

Beckett, A. (2018). Post-Work: The Radical Idea of a World Without Jobs. *The Guardian*.

Bell, P., & Cleaver, H. (1982). Marx's Crisis Theory as a Theory of Class Struggle. *Research in Political Economy, 5*(5), 189–261.

Bhattacharya, T. (2017). *Social Reproduction Theory*. London: Pluto Press.

Boltanski, L., & Chiapello, E. (2005). The New Spirit of Capitalism. *International Journal of Politics, Culture, and Society, 18*(3–4), 161–188.

Bonefeld, W., Gunn, R., & Psychopedis, K. (1992). *Open Marxism* (Vol. 1). London: Pluto Press.

Burnham, P. (1994). Open Marxism and Vulgar International Political Economy. *Review of International Political Economy, 1*(2), 221–231. https://doi.org/10.1080/09692299408434277.

Cederström, C., & Spicer, A. (2015). *The Wellness Syndrome*. Cambridge: Polity Press.

Critchley, S. (2009). Being and Time, Part 5: Anxiety. *The Guardian*.

Fleming, P. (2017). *The Death of Homo Economicus*. London: Pluto Press.

Giddens, A. (1993). *The Third Way: The Renewal of Social Democracy*. London: John Wiley and Sons.

Heidegger, M. (1996). *Being and Time: A Translation of Sein und Zeit*. New York: SUNY Press.

Labour Party. (2017). *Alternative Models of Ownership*. London, UK.

Marx, K. (1970). *Critique of Hegel's "Philosophy of the Right"*. Cambridge: Press Syndicate of Cambridge University Press.

Marx, K. (1999). *Capital*. Oxford: Oxford University Press.

Marx, K. (2004). A Contribution to the Critique of Political Economy. In *Marx Today* (pp. 91–94). New York: Palgrave Macmillan.

Marx, K. (2005). *Grundrisse: Foundations of the Critique of Political Economy*. New York: Penguin Books.

Marx, K. (2007). *Capitalism* (Vol. III). New York: Cosimo Press.

Mercier, A. (2010). Interview with Christian Arnsperger: Capitalism is Experiencing an Existential Crisis. *Truthout*, 15 April.

Mirowski, P. (2013). *Never Let a Serious Crisis Go to Waste*. London: Verso.

Ong, A. (2007). Neoliberalism as a Mobile Technology. *Transactions of the Institute of British Geographers, 32*(1), 3–8. https://doi.org/10.1111/j.1475-5661.2007.00234.x.

Quiggin, J. (2017). Socialism with a Spine: The Only 21st Century Alternative. *The Guardian*.

Sartre, J. (1963). *Search for a Method*. New York: Alfred A. Knopf.

Savall, H., Péron, M., Zardet, V., & Bonnet, M. (2017). *Socially Responsible Capitalism and Management*. London: Routledge.

Soros, G. (2008). False Ideology at the Heart of the Financial Crisis. *The Financial Times*.

Srnicek, N. (2016). *Platform Capitalism*. London: Polity Press.

4

The Facticities of Neoliberalism: Demanding Existential Freedom

The free market arose from a deep-seated disappointment with the post-war liberal consensus. The Keynesian promise to control an unpredictable market had transformed into a society that felt to a growing many as if it was ruled by bureaucracy and a hapless but over-reaching government. While economic reasons are often given for the return of laissez-faire economics, it was in fact existential crisis of post-war Liberalism that ultimately led to its downfall (Krippner 2012; Stein 2011). The truths of a public sector being able to provide for mass prosperity and shape the market were suddenly dispelled by stark economic realities. It led to what then US President Carter called a profound "crises in confidence" (Lipset and Schneider 1983).

The key transition, in this respect, was the transformation of this strong economic Liberalism as a system that opened up social possibilities and expanded humanity's existential freedom, to one that appeared to limit it. In its place was a confident discourse of the free market that revived the sense that both collectively and individually the ability to shape our destiny was once again within our grasp. Consequently, it was put forward, even if only implicitly, as a dramatic existential choice between accepting human limitations or once more embracing the freedom to aspire to

© The Author(s) 2018   **65**
P. Bloom, *The Bad Faith in the Free Market,*
https://doi.org/10.1007/978-3-319-76502-0_4

greater things. This optimism was reflected in President Reagan's 1984 State of the Union address where he proclaimed to the American people and the world that

> Let us begin by challenging our conventional wisdom. There are no constraints on the human mind, no walls around the human spirit, no barriers to our progress except those we ourselves erect. Already, pushing down tax rates has freed our economy to vault forward to record growth.

As the last chapter revealed, it is now the free market that is assailed by its own existential "crises of confidence". The belief that with enough determination an individual could do and achieve anything has been rudely dispelled. Rather the function of the free market is to inform people as to what they cannot do despite their desires otherwise. It is the age of the need for fiscal responsibility, pragmatism, and austerity. An era when experts and leaders are expected to be serious enough to tell people that their dreams are simply too expensive and that desires for a more perfect world are merely wishful thinking. The previously vaunted "third way" has become in time a depressing neoliberal dead end.

The free market is now beholden to supposedly incontrovertible "facts" limiting the social, economic, and political possibilities. It has evolved from an inspiring vision of unrestrained economic growth and individual social mobility to a depressive discourse that cost checks any and all potential for human experimentation and flourishing. It portrays demands for public healthcare, education, and social investment as nice in principle but unfortunately impossible in practice. Hence, Hillary Clinton scoffed at Bernie Sanders's plans to do so as equivalent to promising "to give everyone a pony". It disciplines people to be "pragmatic", realistic about what can be done and circumspect regarding their longings for social or personal change. Hence, the sacred economic laws of the free market have been steadily replaced in importance with the staid but just us unmovable fiscal facts.

Reflected, in turn, is the new hegemony of market freedom. It is less a science and more a shared "common sense" that infuses all contemporary decision-making. Quoting at length the brilliant though sadly recently departed British social theorist Doreen Massey (2013: N.P.):

One of the things that most frustrates me about the current moment is that while there has been a catastrophic crisis of the dominant, neoliberal economic model, in general public political debate there has been no serious challenge to the political and ideological consensus that supports that model. A whole raft of assumptions, running so deep we forget even that they *are* assumptions, remains stubbornly in place. There is, in other words, a hegemonic 'common sense'. It is a common sense of market relations, of competitive individualism, of private gain, the denigration of 'the public', and much else besides. It is a common sense that invades our imaginations and moulds our senses of ourselves.

Even while the reified tenets of classical economics are challenged, their daily influence remains. They exist in the pervasive logics and truisms that define how we presently frame, plan, and act in our everyday lives. It is found in the continued emphasis placed on the need for competition linked with the "universal" truth that it is motivating and breeds innovation. It is discovered in the moral belief that we deserve our debts and must as a society and individually do all we can to repay them (Lazzarato 2009; Muehlebach 2012).

Significantly, this hegemony of market freedom, its "common sense" quality, is fundamental in setting the ultimate limits of social possibility and freedom within a given time and place. It is returning to a theme introduced above, the "facts" that constrain what can and cannot be realistically done. It creates clear boundaries as to which dreams are worth pursuing and which are mere fantasies. In this respect, it confines freedom to specified hegemonic boundaries and in the process marginalizing existential freedom in favour of being "sensible". These "facts" express our essence as humans, a shared understanding of what it means to exist and a common assumption of what is therefore possible. It is the old canard that "Communism is great in theory but impossible in practice due to our human nature".

However, this hegemony is necessarily incomplete—it can never fully extinguish existential desires for a more fundamental freedom. Its essence is always challenged, even if only on the social edges, by different ways to exist and make sense of the world. These "facts" over time become less constricting and more oppressive, clear signs as to the limitations of the present system. As such, they catalyse renewed longings for going beyond these restrictions. They pose fresh challenges for figuring out how to expand human agency. Consequently, they serve as what Sartre would call "facticities", descriptions of existence as it presently is and points of contestation for exploring what existence can potentially become. Thus the "facts" of neoliberalism are transforming into existential "facticities" linked to demands for something much greater than what it currently has to offer.

The Troubling Facts of Neoliberalism

The theory most associated with the free market is neoliberalism. While exact definitions of this term often vary, it broadly refers to the marketization, privatization, and financialization of both the economy and society at large. To this end,

> [n]eoliberalism is in the first instance a theory of political economic practices that proposes that human well-being can best be advanced by liberating individual entrepreneurial freedoms and skills within an institutional framework characterized by strong private property rights, free markets, and free trade. (Harvey 2005: 2)

It therefore justifies free market values in all areas of existence—from government to interpersonal relationships. Of equal importance, it limits the freedom of human action to reflect these market prerogatives.

Traditionally, capitalism was buoyed by a Liberal ideology that reinforced its supposed basis in objective natural laws. Indeed these were grounded in evolutionary theories of "the survival of the fittest", ideas which came, in turn, to legitimize gross inequality (see Hawkins 1997; Hofstadter 1944; Moscovici et al. 1976). Neoliberalism has to a certain extent abandoned such naturalization. It certainly supports neoclassical economics and its objective economic laws. Yet it also believes these market values as having to be culturally promoted to an often irrational and unknowledgeable population.

In this spirit, neoliberalism has made the free market into a present-day common sense. It reflects time-honoured moral truths that are meant to guide individual and collective behaviour. "Neoliberalism is grounded", according to Hall (2011: 10–11),

> in the 'free, possessive individual', with the state cast as tyrannical and oppressive. The welfare state, in particular, is the arch enemy of freedom. The state must never govern society, dictate to free individuals how to dispose of their private property, regulate a free market economy or interfere with the God-given right to make profits and amass personal wealth. State-led "social engineering" must never prevail over corporate and private interests. It must not intervene in the "natural" mechanisms of the free market, or take as its objective the amelioration of freemarket capitalism's propensity to create inequality.

Whereas the overall economics of the free market has been profoundly thrown into question, its ethics have been for the most part popularly preserved. It is still considered worthwhile to be opportunistic, to repay one's debts, to find ever new ways to save time and turn a profit.

This rather "sensible" application of market value is arguably most visible in the moral and political limits it seeks to place in human action. Notably, societies are encouraged at all times to be "fiscally responsible". They must restrain their irresponsible wishes for an easier less oppressive existence for the hard reality of living within their fiscal means. It represents, consequently,

> a new market-centric 'politics'—struggles over political authority that share a particular ideological centre or, in other words, are underpinned by an unquestioned 'common sense'. On the elite level, neo-liberal politics is bounded by certain notions about the state's responsibilities (to unleash market forces wherever possible) and the locus of state authority (to limit the reach of political decision-making. (Mudge 2008: 703)

Personally, people must take moral responsibility for their own lack of marketability. In the face of a near total economic collapse, furthermore, they have a duty to resist urges for selfish and unaffordable public spending and instead accept the mature shared sacrifices of austerity. Thus,

...."austerity" means something rational and commonsensical, something that benefits our life instead of ruining it, something that entails not our sacrifice but our self interest, then we've got a truly moral meaning ... To produce income for oneself and one's family, to plan ahead and budget, to borrow only when it makes sense to do so (but not in order to consume beyond one's current or near-future means), and to save some income, are all acts that are rational, selfish, and thus moral ... To live this way used to be called "economizing." It was the life of the virtuous—and prosperous. (Salsman 2012: N.P.)

The once attractive possibilities of market freedom have been steadily exchanged for a contemporary appeal for people to be pragmatic and sensible. It is an updating of the Protestant ethics for a new age. The demands for greater public welfare are mere fanciful longings, the pipe dreams of the uninformed masses. Those with even a modicum of good sense and education will immediately understand that such desired are simply too expensive to be sustainable.

Here the underpinning logics of neoliberalism merely reflect an unfortunate but unavoidable reality. It would be amazing, it is intoned, to be more collaborative; however, it is only through competing that people are properly motivated to do their best (see Franken and Brown 1995). Socialism would be great but only profit incentivizes people to work and achieve great things. Thus while there is a growing amount of research from a wide range of disciplines such as psychology to neuroscience (see, for instance, Deci et al. 1981) that financial incentives are not the only or even primary motivators for individuals, there is a persistent social discourse to the contrary. A BBC article challenging these ideas, nevertheless, observes that

While opponents of giant rewards to bankers and business executives tend to argue they are unfair on ethical grounds, those in favour usually shift the debate away from morality and towards more pragmatic ground, arguing they are necessary to attract, retain and incentivise talented staff. (Piekema 2014: N.P.)

Similarly, popular wisdom proclaims that "We would love to pay higher wages yet we simply cannot afford it if we want to financially survive". Hence, market freedom is transformed into the liberty to responsibly manage one's desires.

It also means accepting the often hard to sustain rationalizations justifying these market-friendly facts. It is completely unadvisable to ask the wealthy to contribute more in taxes for the public good as this will depress investment an economic growth, no matter that all empirical evidence suggests otherwise. Or it might sound nice to raise the minimum wage but in the long run this will just cause more unemployment—a truth that has been progressively disproved. These are just two examples of how pervasive the "common sense" of the free market remains and how these neoliberal "facts" act as an increasingly dramatic constraint on modern existential freedom.

Facticities Not Facts

Neoliberalism thrives off the spreading of its daily truths. It is a pervasive common sense informing people of what is and is not possible in every area of their existence. The constant appeal to fiscal responsibility frames social reality through overdetermining its accepted limits. They are the iron-clad timeless facts that shape human action and freedom.

A strange truth about incontrovertible universal facts though is that they have a tendency to change dramatically over time. In the case of neoliberalism, they are undermined by the collective weight of personal experience. Indeed, people are seeing that the facts of the free market simply did not add up. For decades, economic growth was promoted as the ultimate measure of human prosperity. Everything else—welfare, employment, personal well-being—was subsumed by and depended on whether they contributed to an expanding economy. The director of the influential pro-capitalist think tank Mark Littlewood noted that "Free marketeers have been gobsmacked. Things we thought of as like the laws of gravity are now up for grabs" (quoted in Beckett 2017: N.P.).

Within organizations, the bottom line was similarly prioritized above all else. However, these reified indicators of progress were challenged by the lived realities that these numbers so often ignored. Thus by 2017, it was duly observed even within the financial press that "Fat profit margins have lifted global stock markets to all-time highs yet wage growth remains well-below average levels, giving credence to the old investment adage that what is good for shareholders is bad for employees" (Subhedar 2017: N.P.).

These profit-based metrics have still retained their status as strict gate-keepers for judging the morality and permissibility of desired actions. Anything was permitted so long as it did not violate these sacred tenets of growth and profitability. It followed a persuasive neoliberal logic that any efforts to decrease these economic gains for those on the top would necessarily depress growth. As reported in a recent *New York Times* article, such economic logic rhetorically asked:

> Why work or invest more if the government will keep almost all the fruits of your troubles? Even Arthur M. Okun, who had been President Lyndon B. Johnson's chief economic adviser, was writing about leaky buckets to illustrate a trade-off between efficiency and equity: Taxing the rich to pay for programs for the poor could slow growth down, in part by reducing the incentive of the rich to earn more. (Porter 2017: N.P.)

They also served as brakes to instituting alternative policies mitigating the effects of this neoliberal order. In this sense, they acted as correctives to "irresponsible" ideas and practices.

However, in the wake of the financial crisis, corporate globalization suddenly seemed not so inevitable. The free market was reframed as an elite myth that profited the rich at the expense of the poor. The calls for a political revolution revealed that people on all sides of the ideological spectrum wanted to renegotiate a new economic deal for the twenty-first century. Even the staunchly pro-neoliberal IMF admitted "Higher tax cuts for rich will cut inequality without hitting growth" (Elliott and Stewart 2017).

This shift in the terms of debate—and the underlying questioning of previously held truths—gesture towards Sartre's concept of facticities. Reinterpreting the ideas of among others Heidegger, he defines facticities as the concrete conditions defining a person's current existence. It is quite profoundly our "being there in the world". This can include where one is born, their physical attributes, their mental gifts, and their relative social privilege. Crucially, they are not who one essentially and permanently is but rather only a present description of who they are now. According to Sartre, "the double property of the human being" is that they are "at once a facticity and a transcendence" (Sartre 1956: 98).

This theory has profound implications for the understanding and application of truth. Facticities represent the present limitations of freedom. Sartre states, thus, that

> The for-itself is, in the manner of an event, in the sense in which I can say that Philip II has been, that my friend Pierre is or exists. The for-itself is, in so far as it appears in a condition which it has not chosen, as Pierre is a French bourgeois in 1942, as Schmitt was a Berlin worker in 1870; it is in so far as it is thrown into a world and abandoned in a "situation" it is as pure contingency inasmuch as for it as for things in the world, as for this wall, this tree, this cup, the original question can be posited: "Why is this being exactly such and not otherwise?" It is in so far as there is in it something of which it is not the foundation—its presence to the world. (Ibid.: 127)

In diagnosing the current constraints on agency and capability they also point the way towards their future opportunities—that which is not there. According to Sartre

> This apprehension of being as a lack of being in the face of being is first a comprehension on the part of the cogito of its own contingency. 'I think, therefore I am'. 'What am I?' A being which is not its own foundation, which qua being, could be other than it is to the extent that it does not account for its being. (Ibid.: 127–128)

These facticities thus stand paradoxically as the limit and the doorway to the creation new forms of freedom. Hence, in Sartre's view,

> It is impossible to grasp faetidty [sic] in its brute nudity, since all that we will find of it is already recovered and freely constructed. The simple fact "of being there," at that table, in that chair is already the pure object of a limiting-concept and as such can not be grasped. Yet it is contained in my "consciousness of being-there," as its full contingency [sic], as the nihilated in-itself on the basis of which the for-itself produces itself as consciousness of being there. The for-itself looking deep into itself as the consciousness of being there will never discover anything in itself but motivations; that is, it will be perpetually referred to itself and its constant freedom. (I am there in order to … etc.) (Ibid.: 132)

Revealed, in turn, is the hegemony of freedom and the potential for its existential resistance.

The Hegemony of Freedom

Freedom is conventionally understood as being challenging to or in negotiation with existent social norms. It has an emancipatory rather than regulative spirit (though it is not necessarily against regulation). Yet the history and present of freedom are deeply rooted in dominant cultural and political values and ideals. Freedoms are enshrined in written constitutions and embedded in shared lived experience. There are also clear social determinations as to what forms of freedom constitute a right and which are at best mere privileges. At its core, the articulation and instantiation of what it means to be free is necessarily hegemonic.

Freedom has always been confronted by a profound paradox. To this end, freedom must be socially established and yet is often justified to be ultimately natural and inherent. The role of government and norms, in this respect, are posited merely to protect those freedoms that are assumed to be innately given by a divine creator or as inherently endowed by nature. At the most basic level, there is a fundamental contradiction to freedom—in that it can never be totally free from its social context. Market freedom, hence, is presumed to be natural or at the very least a matter of "common sense", however, it is in fact a relic of its historical circumstances, emerging within a particular time and place.

There is also an existential paradox afflicting freedom. The practice of a specific freedom is simultaneously an expression and annihilation of existential freedom. Its establishment permits for one type of existence while shutting off the possibility of another from coming into being. Freedom is then always partial—both opening up certain pathways of agencies while shutting off others. Its hegemonic character is hence twofold. First, in its making of one form of agency dominant and then in doing so revealing how this dominant freedom is a limit to social possibility that must be overcome. The modern philosopher J. Melvin Woody (1998: 78), thus, observes that

Freedom may therefore be defined as self-determination, while self-realization may be recognized as a mere stage in the process of self-realization, a process that begins in the ability to entertain possibilities, proceeds to the choice of which possibility to realize, and reaches completion in the absence of constraint in the realization of that chosen possibility.

It is necessary then to understand freedom as a social discourse, shaping our views, actions, and aspirations—the means and ends upon which we "entertain", "choose", and seek to "realize" our possibilities. It is at once regulative and productive, creating the frameworks and socialized avenues to enact and experience agency as well as creativity. Laclau and Mouffe compare discursive hegemony to the rules of a football match. They observe:

> Now, turning to the term discourse itself, we use it to emphasize the fact that every social configuration is meaningful. If I kick a spherical object in the street or if I kick a ball in a football match, the physical fact is the same, but its meaning is different. The object is a football only to the extent that it establishes a system of relations with other objects, and these relations are not given by the mere referential materiality of the objects, but are, rather, socially constructed. This systematic set of relations is what we call discourse. (Laclau and Mouffe 1987: 82)

Hegemony, then, can be defined as "a space in which bursts forth a whole conception of the social based upon an intelligibility which reduces its distinct moments to the interiority of a closed paradigm" (Laclau and Mouffe 1986: 93).

To an extent this may sound like the total negation of freedom, in so much as it sets the literal and figurative boundaries for how one can appropriately behave and interact. Yet it also creates the very social space for freedom to exist, providing for a huge degree of experimentation, empowerment, and even innovation. It is an attempt to "weave together different strands of discourse in an effort to dominate or structure a field of meaning, thus fixing the identities of objects and practices in a particular way" (Howarth 2010: 102). Such an undertaking, whose end goal is often domination, ironically also reveals the contingency and therefore the changeability of our material world and its social systems.

Critically, while such hegemony is discursive, it is not purely rational or meaningful. Rather it encompasses the entire experience of existence. It involves and shapes, to this extent, what is "sensible" not only morally but also sensually. Gramsci's "common sense", therefore, points to the unifying aspect of "common sense". It is quite literally our shared understandings as well as shared ways for emotionally and physically engaging reality. This common sense extends to the experience of freedom, as how people dominantly embody agency becomes a cornerstone for shaping their interactions and relationships.

Market freedom is then the epitome of a modern "common sense". It stands as a hegemonic framework for pursuing one's desires. It is the rules of the contemporary social game for making decisions and planning action. It channels individual and collective agency to reflect its version of reality. Significantly, it represents a shared form of sense-making, binding individuals into a communal emotional and sensual experience. It makes its existence universally relatable, one found in the anxiety of seeking employment, the struggles associated with social mobility, the difficulties of balancing work and life. Just as critically, it distinguishes who is free and who is not and to what degree. It also creates common physical spaces to embody this freedom (or relative lack thereof)—from the noisy factory to the contemporary open plan office. Yet it is precisely, this shared sense of being in the world that gestures towards the possibility of another existence, one beyond the current limits of "sensible" freedom.

Articulating Freedom

Existentialism is rather famous for being action driven. To paraphrase one of its most known ideas "you are your choices". By contrast, hegemony is largely associated with the discursive, the norms and ideas that are thought to shape such action. It would seem at first appearance that existentialism and hegemony are at best strange theoretical bedfellows and at worst completely incompatible. However, as outlined above, what constitutes freedom and therefore the framework of one's choices is necessarily socially constructed. Yet, equally important such socialization can never exhaust the potential for freedom or capture the full possibilities of our existence. What is key, in this regard, is the articulation of freedom as a means for concretely discovering and choosing to transcend its existing limits.

A crucial but nevertheless often unasked question is how do people know what it means to be free? How do they find out what freedoms they supposedly have in theory and practice? Tellingly, people must be told how to be free, informed to an extent of their agency as social subjects. This is apparent in neoliberalism, for instance, in the need to teach individuals and communities how to be "entrepreneurial". For this purpose, neoliberalism

> advocates a programme of deliberate intervention by government in order to encourage particular types of entrepreneurial, competitive and commercial behaviour in its citizens, ultimately arguing for the management of populations with the aim of cultivating the type of individualistic, competitive, acquisitive and entrepreneurial behaviour which the liberal tradition has historically assumed to be the natural condition of civilised humanity, undistorted by government intervention. (Gilbert 2013: 9)

It is, hence, a type of moral education for the general population—a contemporary civilizing process aimed at having them embrace ethical advice about being entrepreneurial, efficient, and productive. The canard that "these freedoms are self-evident" takes on, in turn, a new meaning. It is when freedoms that are culturally derived become "common sense", so obvious that they appear fully rational and inevitable, that they are at their strongest.

However, it is in the articulation of these freedoms that they are also at their most socially exposed and vulnerable. More precisely, in having to define what freedom is, it is also revealed what it is not or even more accurately what it is not yet. There is an important philosophical movement to make here, in the shifting of focus from consciousness to articulation. Sartre tellingly prioritizes the ways consciousness informs the relation of Being and Nothingness, in so much that it is in our awareness of our existence (e.g. "Being-in-itself) that we become equally aware of non-being, that which is not there and missing. Quoting him at length on the subject

Facticity is not then a substance of which the for-itself would be the attribute and which would produce thought without exhausting itself in that very production. It simply resides in the for-itself as a memory of being, as its unjustifiable presence in the world. Being-in-itself 'can found' its nothingness but not its being. In its decompression it nihilates itself in a for-itself which becomes qua for-itself its own foundation; but the contingency which the for-itself has derived from the in-itself remains out of reach. It is what remains of the in-itself in the for-itself as facticity and what causes the for-itself to have only a factual necessity; that is, it is the foundation of its consciousness-of-being or existence, but on no account can it found its presence. Thus consciousness can in no case prevent itself from being and yet it is totally responsible for its being. (Sartre 1956: 133)

Yet as this implies consciousness is never direct, it is necessarily grounded in an embedded context, shaped by the facticities in which it is born into and grows up within. Central to this process of conscious development is the articulation of reality, the shared "common sense" that shapes how we perceive our existence. And it is precisely, in this articulation of Being, that non-being also emerges, making us aware both of what is now and what is not but could be.

This encounter with nothingness, with that which does not exist, is ultimately much more than a matter of articulation—its mere public utterance. This lack is fundamentally experienced, for as Sartre is apt to remind us

Thus the possible can not be reduced to a subjective reality. Neither is it prior to the real or to the true. It is a concrete property of already existing realities. In order for the rain to be possible, there must be clouds in the sky. To suppress being in order to establish the possible in its purity is an absurd attempt. The frequently cited passage from not-being to being via possibility does not correspond to the real. To be sure, the possible state does not exist yet; but it is the possible state of a certain existent which sustains by its being the possibility and the non-being of its future state. (Ibid.: 150)

This existential gap though is concretely manifested in two distinguishable but ultimately inter-related ways. The first is a specific absence of an existent freedom, the feeling, and concrete realization that others have more of a certain type of agency than yourself. The second is a fundamental experience of the limits of this freedom, the awareness that these social forms of empowerment are in fact limitations on what one can possibly do. Consequently

> although brute things (what Heidegger calls "brute existents") can from the start limit our freedom of action, it is our freedom itself which must first constitute the framework, the technique, and the ends in relation to which they will manifest themselves as limits. Even if the crag is revealed as "too difficult to climb," and if we must give up the ascent, let us note that the crag is revealed as such only because it was originally grasped as "climbable"; it is therefore our freedom which constitutes the limits which it will subsequently encounter. (Ibid.: 482)

It is this interjection of possibility that separates this experience of limitation with a simple restriction. It is not a legal or proscriptive barrier to one's liberty, the prohibition, for instance, on murdering someone. Instead, it is the awareness that even at its most perfect a current freedom is in fact limited as to what it can allow humans to individually and collectively accomplish. Thus, for example, market freedom once heralded for its sole capacity for providing people with social mobility is increasingly seen as unable to deliver shared prosperity or sustainable social, economic, or political progress. Accordingly, it is the experience of nothingness, not merely an identifiable lack.

Arising, therefore, is the articulation of a deeper existential freedom. The rendering of Being as contingent rather than necessary or complete. It is the expression of a shared demand for fresh means to shape the present and future. Revealed is the incompleteness of prevailing freedoms and the demand to explore what is currently not existent for potential alternatives. While often visionary, it is not purely abstract or conceptual. By contrast, it is the return of existence over essence, where people experiment with novel ideas and practices in the quest to discover their Being anew.

The Facticities of Neoliberalism

The free market is conventionally framed as being essentially enabling. Values of competition and capitalist reward create an inherent incentive for pursuing one's aspirations and developing one's talents. At its most ideal, it represents a freedom from the tyranny of government and cultural privilege. Collectively it points to a romantic vision of a society constantly pushed forward by capitalist innovation.

It enables thus both personal betterment and shared progress.

Yet in its actual realization it brings to the fore all the things it cannot control or achieve. Specifically, issues of inequality, poverty, crumbling infrastructure appear endemic and increasingly permanent. Neoliberal ideologues and policy-makers, in turn, adopt a policy of blame. It is the fault of backward and corrupt governments who refuse to properly implement this free market agenda. Consequently

> From the early 1990s onwards, the call for less state has gradually been substituted by a call for a better state. This new approach should not be confused with a plea for a return to the strong (Keynesian or socialist) state. Rather it implies better and transparent governance of what is left of the state after neoliberal restructuring has been implemented. (Demmers et al. 2004: 2)

This produced, in turn, the rise of "self-disciplining capitalist states" in which

> The overriding emphasis is one of maintaining capitalist discipline, not giving into the lures of corruption or the populist but ill-conceived desires of many of its citizens. Governments, in this respect, have taken on an almost paternalistic role. Forcing their populations to be "responsible" and "selfdisciplined" as if the acceptance of neoliberal reforms was a matter of personal and national maturity. To ensure this "mature" perspective states must be willing to act decisively and at times coercively to stamp out the threat of "irresponsibility". (Bloom 2016: 141)

Or it is people themselves who are to blame for being lazy or irrational. "Never mind structural unemployment: if you don't have a job it's because you are unenterprising", notes social critic George Monbiot (2016: N.D.).

Never mind the impossible costs of housing: if your credit card is maxed out, you're feckless and improvident. Never mind that your children no longer have a school playing field: if they get fat, it's your fault. In a world governed by competition, those who fall behind become defined and self-defined as losers.

When all else fails, neoliberalism embracese a tragic essentialism. It is simply human nature—inequality and privilege will always persist in some form or another.

This embrace of essentialism reflects how Being comes to supersede and displace existence. Put differently, that the possibilities of existing become discursively confined to a single particular way of Being. This is witnessed in the discourse of inevitability accompanying corporate globalization and recently technological transformation. In this respect, it "naturalises the market and the economy, to such an extent that it presents the latter as [an] autonomous force to which we must bow" (De Angelis 1997: 43).

It also represents an ironic shift in the portrayal of market freedom. It was now a tool for combatting these endemic social ills. Accordingly, "good governance is not about building roads or transmitting knowledge, it is about establishing systems of management and self-control to increase 'efficiency' and 'transparency' by means of auditing and new public management" (Anders 2010: 165). The emphasis on fiscal innovation became an imperative for social reform innovatively using financial ideas. Hence, a recent large-scale transnational survey revealed

> that economic elites do feel a sense of social responsibility yet linked to continuing profit and fiscal sustainability. The how of this responsibility differed between countries, yet the problematic remained the same. Similar governmentality for governments, how to ensure social fiscal responsibility. (Witt and Redding 2011: 59)

This mixture of fiscal rationalities with social idealism can lead to quite forward-thinking places—ones that may challenge elite financial interests even as it uses its very language. Holzmann and Jorgensen (1999) recommend linking social protection to risk assessment in a manner that "emphasizes the double role of risk management instruments—protecting basic livelihood as well as promoting risk taking" (529). Similarly, some scholars argue for introducing a "pro-poor (public) expenditure index" that would serve as a conditionality for countries receiving further aid while granting donors "greater flexibility" in auditing and punishing those who fail to keep these commitments (Mosley et al. 2004).

Ideas of social entrepreneurship illustrate the efforts to expand the limits of the free market without sacrificing its core components (Cho 2006; Peredo and McLean 2006; Seelos and Mair 2005). It plays on growing popular desires for "ethical capitalism" (Barry 2004) and "green capitalism" (Smith 2018; Tienhaara 2013; Watts 2002). Emerging alongside this more moral capitalist was "a state that performs the role of a market actor and regulator at the same time" (Livne and Yonay 2015). In particular, governments were now tasked with "managing" this attempted balance between broader public welfare and economic growth (Kaul et al. 2003). This universal aspiration, moreover, had to be locally implemented as states were expected to work within and counteract their specific cultural conditions to make neoliberalism socially responsible (Engelen et al. 2008; Kalinowski 2012). Their overall purpose was to put into place "national and regional experiments with global ideas" (Roy et al. 2006).

Theoretically, this points to what Laclau and Mouffe refer to as a logic of difference (1986). Here a hegemonic discourse incorporates previously antagonistic beliefs and practices. While certainly co-optive, it is also a quite transformative moment. Specifically, it articulates that a given status quo is limited and as such must be altered and enlarged. Existentially, it repurposes existent freedoms towards new and often different social ends. In doing so, it tenders these freedoms partially contingent—their survival linked to the agency they provide people to address demands for change.

These put into question, furthermore, the facts sustaining these freedoms. Suddenly, responsibility is extended beyond the mere individual to the corporation. Moreover, while corporate social responsibility (CSR) is clearly dedicated to implementing half measures there lies a more radical truth at its heart. It is that past fiscal constraints must contend or at least be negotiated with a broader social justice mission. Its attempts and failures to do so cumulatively begin to reveal the limits of market freedom and the need to go beyond them. Hence, the facts of the free market are gradually transforming into the facticities of neoliberalism. They are the conditions of a particular mode of capitalist Being—slowly recognized as being historical rather than essential. The expectations of existing freedoms are widened and partially redirected. And these facticities serve as a springboard for politically demanding more not only from existent market freedom but also of the social possibilities of freedom itself.

Demanding Existential Freedom

Neoliberalism is facing an existential crisis as discussed in the previous chapter. Its assumptions and worldview are under cultural attack. It has evolved from a radical discourse promising the expansion of human agency to restrictive ideology holding back personal and collective freedom. It represents a hegemony of freedom, a dominant and naturalized way of acting and pursuing one's desires. Yet there are growing demands to enlarge the scope of freedom away from the limits imposed on it by the free market.

Freedom is popularly linked to the imagination. It is the freedom to conceive a different world, an alternative existence whether in a parallel universe or a not yet arrived future. It is telling that imagination is so commonly associated with escapism—indicative of a desire to experience, even if only in our fantasies, another way of Being. It also points to something perhaps much more radical. Escapism is the mere symptom of a society plagued by a chronic and seemingly fatal lack of existential freedom.

In his later work, Laclau refers to this condition, perhaps not coincidentally, as a "social imaginary" referring to "a horizon: it is not one among other objects but an absolute limit which structures a field of intelligibility and is thus the conditions of possibility for the emergence of any object" (1991: 63). It represents, thus, the discursive horizon of possibility—the limits of what can be realistically conceived within a given time and place. Within this horizon, there may be quite a diverse array of options. Using the example of Peronist Argentina, Thomassen argues that

If the imaginary horizon—including the central antagonism—of Argentine politics was the same for almost half a century, then it seems wrong to talk about instability. There was stability with regard to the terms of political and social struggles, and there was a common imaginary shared by the whole political order, including the antagonistic forces. (2005: 299)

However, these are only ever options. The genuine choice is ultimately constrained to what is socially imaginable. It is the atmosphere of freedom, its cultural gravity anchoring people as if by some invisible natural law to a particular hegemonic grounding of what is or is not sensible.

Specific to the free market, the underlying tenants of neoliberalism become the very conditions of possibility for existence as such. Quoting the social theorist Mark Fisher (2010)

In the 1960s and 1970s, capitalism had to face the problem of how to contain and absorb energies from outside. It now, in fact, has the opposite problem; having all too successfully incorporated externality, how can it function without an outside it can colonize and appropriate? For most people under twenty in Europe and North America, the lack of alternatives to capitalism is no longer even an issue. Capitalism seamlessly occupies the horizons of the thinkable. Jameson used to report in horror about the ways that capitalism had seeped into the very unconscious; now, the fact that capitalism has colonized the dreaming life of the population is so taken for granted that it is no longer worthy of comment.

It culturally delineates, thus, what is "real" and "sensible" and what is mere fantasy.

The metaphor of the imaginary horizon is apt, in this respect. Existential freedom is found in looking up beyond what is given, trying to conceive in one's minds what is not yet physically or socially possible. It is exactly here that social facts transition into existential facticities. It goes from a strict cultural limit to a provocative social challenge. It is a demand for making the currently impossible possible.

Consequently, the existent limits of imagination are rendered questionable, contingent, and utterly replaceable. Importantly, these limits are rearticulated as a barrier to freedom as well as their supporters. Within contemporary neoliberalism, the conventional depiction of financial titans and increasingly their pro-market political allies as "Masters of the Universe" takes on a potentially different meaning. Instead of mere veneration or even resigned acceptance, both their rule and the conception of the universe are profoundly questioned. Using other language, it is not just that they have power but fundamentally a challenge to their narrow conception of present and future human existence.

In such moments a new desire for freedom is activated giving rise to fresh possibilities for Being. Specifically, it speaks for the very logics which guide and set into motion socially possible, referring to Laclau's words,

> the type of relations between entities that make possible the actual operations of that system of rules. While the grammar enunciates what the rules of a particular language are, the logic answers to a different kind of question: how entities have to be to make those rules possible. (Butler et al. 2000: 283–284)

Glynos and Howarth (2008), to this effect, distinguish between social and political logics. The social is associated with a process of naturalization, in which hegemonic beliefs and practices are simply "common sense". By contrast, politics is the denaturalization of these social "truths", opening them up to contestation and dramatic alteration. Added to these is what can be termed a "freedom logic"—which is the outgrowth of politics, the evolution of dislocation and denaturalization to a highlighting of the existential ability to explore, pursue, and realize alternative forms of social existence.

Accordingly the radical political moment is transformed into a revolutionary existential opportunity. Drawing again on Laclau, he links politics to the populist demand. In this respect, it represents

> If the situation remains unchanged for some time, there is an accumulation of unfulfilled demands and an increasing inability of the institutional system to absorb them in a differential way (each in isolation from the others), and an equivalential relation is established between them. The result could easily be, if it is not circumvented by external factors, a widening chasm separating the institutional system from the people. (Laclau 2005: 73–74)

More precisely, a particular demand for change morphs into a universal demand for total transformation when left unaddressed. It is populist in that it creates a chain of equivalence between individuals and communities, unifying them in the desire for this radical social change. Just as significantly, it serves as the foundation for an existential demand for freedom. Specifically, it is the articulated longing to have a fundamental choice regarding the character of our social existence. The "people" develop from their anti-establishment beginnings into a force for reasserting the shared freedom to shape their living reality separate from their existing "naturalized" limitations.

Existential freedom is therefore a revolutionary demand and radically demanding. It is captured, once again, in the shift from "facts" to "facticities". Once a truth is revealed as contingent, a social constraint that can and perhaps must be overcome, there comes with it the responsibility to create the conditions necessary for going beyond this once unquestioned "common sense". Here, the "facticity" becomes radically demanding. How, for instance, can we move beyond market freedoms to ensure greater equality and social inclusion? What would an alternative form of globalization look like and how can it be realized? These profound questions also catalyse the search for new concrete freedom to realize such creative social rethinking. Equally important, the freedom to reimagine our existence brings to the fore ethical considerations of what precisely a shared form of existence should be. Hegemony, accordingly, is always at its core revolving around this play of freedom.

Yet as will be shown, it is exactly this fundamentally demanding quality of freedom that can make it so daunting and, in turn, make the ontological security of an existent status quo so appealing. While the prospect of new possibilities for existing may be exciting, the fear of not existing at all can lead us to surrender our freedom and cling even more strongly to the narrow experience of Being that we know. The survival of the free market, and the death of freedom, is found in the false choice between Capitalism Being and Nothingness.

References

Anders, G. (2010). *In the Shadow of Good Governance: An Ethnography of Civil Service Reform in Africa (Afrika-Studiecentrum Series, v. 16)*. Brill Academic Publishers.

Barry, A. (2004). Ethical Capitalism. In W. Larner & W. Walters (Eds.), *Global Governmentality: Governing International* (p. 195). London: Routledge.

Beckett, A. (2017). How Britain Fell Out of Love with the Free Market. *The Guardian*.

Bloom, P. (2016). *Authoritarian Capitalism in the Age of Globalization*. Cheltenham: Edward Elgar.

Butler, J., Laclau, E., & Žižek, S. (2000). *Contingency, Hegemony, Universality: Contemporary Dialogues on the Left*. London, New York: Verso.

Cho, A. (2006). Politics, Values and Social Entrepreneurship: A Critical Appraisal. In J. Mahair (Ed.), *Social Entrepreneurship* (pp. 34–56). London: Palgrave Macmillan.

De Angelis, M. (1997). The Autonomy of the Economy and Globalisation. *Common Sense, 21*, 41–59.

Deci, E., Nezlek, J., & Sheinman, L. (1981). Characteristics of the Rewarder and Intrinsic Motivation of the Rewardee. *Journal of Personality and Social Psychology, 40*(1), 1–10. https://doi.org/10.1037//0022-3514.40.1.1.

Demmers, J., Fernandez, A., Jilberto, E., & Hogenboom, B. (2004). *Good Governance in the Era of Global Neoliberalism: Conflict and Depolitisation in Latin America, Eastern Europe, Asia, and Africa*. London: Taylor and Francis.

Elliott, L., & Stewart, H. (2017). IMF: Higher Taxes for Rich Will Cut Inequality Without Hitting Growth. *The Guardian*.

Engelen, E., Konings, M., & Fernandez, R. (2008). The Rise of Activist Investors and Patterns of Political Responses: Lessons on Agency. *Socio-Economic Review, 6*(4), 611–636. https://doi.org/10.1093/ser/mwn012.

Fisher, M. (2010). *Capitalist Realism*. Winchester: Zero Books.

Franken, R., & Brown, D. (1995). Why Do People Like Competition? The Motivation for Winning, Putting Forth Effort, Improving One's Performance, Performing Well, Being Instrumental, and Expressing Forceful/Aggressive Behavior. *Personality and Individual Differences, 19*(2), 175–184. https://doi.org/10.1016/0191-8869(95)00035-5.

Gilbert, J. (2013). What Kind of Thing Is 'Neoliberalism'? *New Formations, 80*(80), 7–22. https://doi.org/10.3898/newf.80/81.introduction.2013.

Glynos, J., & Howarth, D. (2008). *Logics of Critical Explanation in Social and Political Theory*. London: Routledge.

Hall, T. (2011). The Triple Bottom Line: What Is It and How Does It Work. *Indiana Business Review, 86*(1), 4–8.

Harvey, D. (2005). *A Brief History of Neoliberalism*. Oxford: Oxford University Press.

Hawkins, M. (1997). *Social Darwinism in European and American Thought, 1860–1945: Nature as Model and Nature as Threat*. New York: Cambridge University Press.

Hofstadter, R. (1944). *Social Darwinism in American Thought*. Boston: Beacon Press.

Holzmann, R., & Jorgensen, S. (1999). Social Protection as Social Risk Management: Conceptual Underpinnings for the Social Protection Sector Strategy Paper. *Journal of International Development, 11*(7), 1005–1027. https://doi.org/10.1002/(sici)1099-1328(199911/12)11:7<1005::aid-jid643>3.0.co;2-b.

Howarth, D. (2010). *Discourse*. Buckingham, UK: Open University Press.

Kalinowski, T. (2012). Regulating International Finance and the Diversity of Capitalism. *Socio-Economic Review, 11*(3), 471–496. https://doi.org/10.1093/ser/mws023.

Kaul, I., Conceicao, P., Le Goulven, K., & Mendoza, R. (2003). *Providing Global Public Goods: Managing Globalization*. New York: Oxford University Press.

Krippner, G. (2012). *Capitalizing on Crisis*. Cambridge: Harvard University Press.

Laclau, E. (1991). *New Reflections on the Revolution of our Time*. London: Verso.

Laclau, E. (2001). *New Reflections on the Revolution of Our Time*. London: Verso.

Laclau, E. (2005). *On Populist Reason*. London: Verso.

Laclau, E., & Mouffe, C. (1986). *Hegemony and Socialist Strategy*. London: Verso.

Laclau, E., & Mouffe, C. (1987). Post-Marxism Without Apologies. *The New Left Review, November–December, 166*, 79–106.

Lazzarato, M. (2009). Neoliberalism in Action. *Theory, Culture & Society, 26*(6), 109–133. https://doi.org/10.1177/0263276409350283.

Lipset, S., & Schneider, W. (1983). *The Confidence Gap*. Baltimore, MD: Johns Hopkins University Press.

Livne, R., & Yonay, Y. (2015). Performing Neoliberal Governmentality: An Ethnography of Financialized Sovereign Debt Management Practices. *Socio-Economic Review, 14*(2), 339–362. https://doi.org/10.1093/ser/mwv019.

Massey, D. (2013). We Need to Challenge the Hegemonic 'Common Sense' of Market Relations, of Competitive Individualism, of Private Gain, the Denigration of 'the Public', and Much Else Besides. *LSE Blog*.

Monbiot, G. (2016). Neoliberalism – The Ideology at the Root of All Our Problems. *The Guardian*.

Moscovici, S., Sherrard, C., & Heinz, G. (1976). *Social Influence and Social Change [engl.]*. New York: Published in cooperation with European Association of Experimental Social Psychology by Academic Press.

Mosley, P., Hudson, J., & Verschoor, A. (2004). Aid, Poverty Reduction and the 'New Conditionality'. *The Economic Journal, 114*(496), F217–F243. https://doi.org/10.1111/j.1468-0297.2004.00220.x.

Mudge, S. (2008). What Is Neoliberalism. *Socio-Economic Review, 6*(4), 703–731.

Muehlebach, A. (2012). *The Moral Neoliberal: Welfare and Citizenship in Italy*. Chicago: The University of Chicago Press.

Peredo, A., & McLean, M. (2006). Social Entrepreneurship: A Critical Review of the Concept. *Journal of World Business, 41*(1), 56–65. https://doi.org/10.1016/j.jwb.2005.10.007.

Piekema, C. (2014). Does Money Really Motivate People? *BBC*.

Porter, E. (2017). Tax Cuts, Sold as Fuel for Growth, Widen Gap Between Rich and Poor. *The New York Times*.

Roy, R., Denzau, A., & Willett, T. (2006). *Neo-Liberalism: National and Regional Experiments with Global Ideas*. London: Routledge.

Salsman, R. (2012, May 17). Fiscal Austerity and Rational Morality. *Forbes*.

Sartre, J. (1956). *Being and Nothingness*. New York: Gallimard.

Seelos, C., & Mair, J. (2005). Social Entrepreneurship: Creating New Business Models to Serve the Poor. *Business Horizons, 48*(3), 241–246. https://doi.org/10.1016/j.bushor.2004.11.006.

Smith, J. (2018). Green Capitalism: The God that Failed. *Real World Economics Review, 56*, 112–144.

Stein, J. (2011). *Pivotal Decade: How the United States Traded Factories for Finance in the Seventies*. New Haven: Yale University Press.

Subhedar, V. (2017). Waiting for Wage Growth: An Eye on Corporate Margins. *Reuters*.

Thomassen, L. (2005). From Antagonisms to Heterogeneity: Discourse Analytical Strategies. *Essex Papers in Politics and Government, Sub-Series in Ideology and Discourse Analysis, 21*, 1–37.

Tienhaara, K. (2013). Varieties of Green Capitalism: Economy and Environment in the Wake of the Global Financial Crisis. *Environmental Politics, 23*(2), 187–204. https://doi.org/10.1080/09644016.2013.821828.

Witt, M., & Redding, G. (2011). The Spirits of Corporate Social Responsibility: Senior Executive Perceptions of the Role of the Firm in Society in Germany, Hong Kong, Japan, South Korea and the USA. *Socio-Economic Review, 10*(1), 109–134. https://doi.org/10.1093/ser/mwr026.

Woody, J. (1998). *Freedom's Embrace*. Pennsylvania: Pennsylvania State University.

5

Capitalist Being and Nothingness: Enjoying Existential Freedom

Capitalism is often associated with being at its heart aspirational. It is meant to above all else fulfil people's deepest desires. Written into its very constitution is the right to "pursue happiness". It is exemplified in the myth of the American Dream. Underpinning the quite regulative everyday existence of capitalism is a utopian vision of limitless person possibility. In the words of noted US economist Robert Reich, "The faith that anyone could move from rags to riches—with enough guts and gumption, hard work and nose to the grindstone—was once at the core of the American Dream".

The twenty-first century has turned these dreams for a growing many into a living nightmare. Far from being aspirational, they now seem to be a sick joke handed down from previous generations. As one prominent newspaper during the 2016 presidential election proclaimed "Bernie Sanders and Donald Trump look like saviours to voters who feel left out of the American Dream" (Sandel 2016: N.P.). It is not surprising then that so much contemporary culture is populated with apocalyptic themes and images. It reflects "a determinate political subjectivation—and the representation of the end of the world so common in these narratives"

© The Author(s) 2018
P. Bloom, *The Bad Faith in the Free Market*,
https://doi.org/10.1007/978-3-319-76502-0_5

(Lanci 2014: 26). The capitalist end of history has transformed into the figurative perhaps even literal end of the world. It represents a type of "zombie capitalism" (see especially Giroux 2011). Hence,

> in the twenty-first century capitalism as a whole is a zombie system, seemingly dead when it comes to achieving human goals and responding to human feelings, but capable of sudden spurts of activity that can cause chaos all around. (Harman 2009: 11)

Contemporary capitalism is infused with a profound sense of nihilism. It can imagine nothing beyond its own narrow beliefs, and as these fade, it increasingly is willing to consign the plant to a future of nothingness. In the modern era, the free market

> congregation believes in neo-liberalism as a social vision, that fair and free competition will allow those who deserve it to advance in the market in the pursuit of happiness by the sweat of their brows. Yet in a bizarre twist of fate the money-lenders have become the priests, and they believe in a different God. Of course they proclaim the market's social value, but what really answers their prayers is the unlimited sanction of the pursuit of self-interest. This is not a moral vision, it is nothing short of neo-liberal nihilism. (Gabriel 2009: N.P.)

It offers people a stark and depressing choice between a steady decline and a rapid disintegration. To paraphrase an old rock n roll cliché, the most pressing question for the modern-day capitalist subject is whether it is better to burn out or fade away?

If all is so hopeless it must be asked then what is the continuing hold that capitalism and the free market have over us? To a certain degree, it is precisely that there is no immediate or obvious alternative available. This echoes Zygmunt Bauman's (2006: 103) contention that "in contemporary dreams … the image of 'progress' seems to have moved from the discourse of *shared improvement* to that of *individual survival*."

It is, in this respect, to believe in something—no matter how horrible—or nothing, that which does not exist and remains utterly unknown even to our wildest imagination. It is a desperate rationality—where capitalism remains the only game in town, the other option being cataclysmic or supernatural ruin.

Yet it also has a more mundane but every bit as significant psychic grip on us. Capitalism thrives on creating and promising to fulfil a person's feelings of lack. Its survival depends on eternal fantasies of wholeness linked to consumption and labour. If only we worked harder, we could get promoted. If I buy this outfit, I will finally truly be happy. Revealed is an ironic reconfiguration of Sartre's famous idea of "Being and Nothingness". For him, this conscious and lived experience of nothingness—that which we are not—created a lack that catalysed our freedom to become something different than our current selves.

Nevertheless, the capitalist lack is much more resilient, subtle, and insidious. It is paradoxically our consciousness of our lack that motivates us to invest even more heavily into the fulfilling possibilities of the market. At stake is the fundamental fantasy of capitalism at work. It is precisely in this eternal attempt to deal with our lack that we are granted a precarious and much desired security as capitalist subjects. The pursuit of this fulfilment covers the "real" truth that we can never be fully sated in our drive for psychic wholeness.

Underpinning this fantasy is a fear that without capitalism and its manageable sense of lack, we would disintegrate into nothingness. Humanity is thus tragically held in thrall by the fantasy of capitalist being and nothingness.

Unfulfilling Capitalism

The financial crisis that has largely defined the new millennium has been viewed by most as a threat rather than an opportunity. Theoretically, these dislocations would be expected to be greeted with if not complete excitement than at least cautious optimism. Even before the near global economic meltdown, the destructive effects of neoliberalism were clearly evident. These economic anxieties were compounded by the apocalyptic dangers of climate change and terrorism.

Indeed, even after the initial shock the crash primarily bred fear—one that has fuelled a diverse set of anti-establishment responses from both the Right and the Left. Specifically,

In Europe and the U.S., the movement is fuelled by middle-class economic insecurities, exacerbated by fear of immigrants arriving to steal jobs or soak up welfare money and other taxpayer dollars. Those anxieties are overlaid with an absolute conviction among many citizens that existing political leaders either don't understand or, worse yet, don't care. (Seib 2016)

Moreover, these pressing concerns reflected deeper longer term fears over the future of the economy and their own place in it. According to Nobel Laureate economist Robert Shiller (2017: N.P.),

My own theory about today's stagnation focuses on growing angst about rapid advances in technologies that could eventually replace many or most of our jobs, possibly fueling massive economic inequality. People might be increasingly reluctant to spend today because they have vague fears about their long-term employability—fears that may not be uppermost in their minds when they answer consumer confidence surveys.

Obviously uncertainty—especially on such a massive scale—is a potent recipe for panic. Yet when the full scale of the problem was exposed and the financial order was revealed to be at best completely corrupted and at worst irredeemably rotten, the outcry for change was matched in intensity by the shared desire for recovery. Indeed, the aftermath of the near global financial crash was a persistent story of "crises and recovery" (see Hayward 2010). This seeming contradiction speaks to the deep psychic attachment that people still have towards the free market.

In part, this focus on recovery represents a broader affective crises narrative. The spectre of complete social breakdown brings with it understandable social anxiety. As the scholars Hicks and Dunn darkly observe "beneath the chatter of crisis and redemption make sure to listen for the slaughtering bell". There emerges for this reason a pronounced nostalgia for the past—especially its once guaranteed promise of a stable and prosperous market future.

At first, it appeared that hope would triumph. The election of Barack Obama signalled the dawning of a new era. The existential moment was exemplified by the chants of "Yes we can" by historically marginalised people of colour and a politicized younger generation. However, these dreams of change were quickly dashed. Around the world, the emphasis was on reforming the status quo instead of replacing it. Thus "States of anxiety, insecurity, outrage, scepticism and more have been mobilized, managed, negotiated, contained and directed with few signs of success" (Clarke 2014: 109).

There is an obvious story to tell in retrospect that the incrementalism of Obama and the conservative embrace of austerity elsewhere was simply an instance of elite retrenchment. However, appearances can also be misleading. Those on the top also preyed on the affective bond that the wider populace has with the free market order, even as those on the bottom increasingly expressed their frustration with it. Namely, the individual futures that they had invested themselves in and the comfortable knowledge of the supposed market freedoms that would let them potentially one day achieve them (Bloom 2014).

Even if rationally, these aspirations were exposed as eternally out of reach, personalized myths of ultimate victory in a system firmly rigged against them—emotionally they are hard to simply abandon. Saving the system, therefore, was an attempt—at times even unconsciously—for preserving one's dreams. This is reflective of the free market fantasy that "the sky's the limit" (Bloom and Cederstrom 2009). To this end, capitalism becomes the exclusive engine of people's desire. It is the sole means through which they can pursue fulfilment. Without it, there is palatable fear that they will be left with nothing.

Being and Nothingness

The fear of nothingness looms large in the present era. In the face of what appears to be utter systematic failure, the question of what comes next is quite ominous indeed. Capitalism may be broken but it is nonetheless a known quantity. By contrast, alternatives that do exist conjure up images of a recently oppressive past or coming soon dystopian future. There seems to be nowhere left to go historically but accept our capitalist fate.

However, for Sartre nothingness is actually fundamental to Being itself. He distinguishes, as discussed previously, between "being-in-itself" and "being-for-itself", the former representing the existence of objects and the latter associated with the human quality of consciousness. For conscious beings, nothingness is absolutely crucial to the experience of Being. In Sartre's view, we are materially born into our body and inserted into a socialized existence. Consciousness, in this respect, has the capacity to articulate and put this being into mental perspective and as such conceive as well as negate certain possibilities. It is able to conceptualize what Being is, what it could be, and just importantly what it is not. Therefore, in Sartre's (1956: 42–43) view

> The not, as an abrupt intuitive discovery, appears as consciousness (of being), consciousness of the not. In a word, if being is everywhere, it is not only Nothingness which … is inconceivable; for negation will never be derived from being. The necessary condition for our saying not is that nonbeing be a perpetual presence in us and outside of us, that nothingness haunt being.

Critically, this deep relationship with non-being is not merely subjective. Instead it is a lived experience of nothingness. It reflects one's embedded existence, their insertion into a given culture and personal circumstance, creating the conditions for exposure to a certain existence and the negation of others. To this end, "it is evident that non-being always appears within the limits of a human expectation" (Ibid.: 38). It is crucial, here, to point out the difference between what can be considered inherently impossible and that which is an example of non-being. Sartre distinguishes concrete from what he terms abstract nothing. Concrete nothingness is a lack, a recognition of non-being, such as "I live in a flat not a house" OR "inequality is high not low". Abstract nonbeing, conversely, is an essential impossibility, like that a triangle does not have four sides.

Humans then are necessarily concretely involved with nothingness. It is a matter of what non-being emerges from their lived existence. To this end, Sartre philosophizes that nothingness constitutes a broader totality that shapes our experience of being. Prominent Sartrian commentator Joseph Catalano uses the example of the blind person, whose lack of sight is part and parcel to a broader totality formed in relation to and revolving around this absence. He notes,

Classically, the term 'not-seeing', is considered a mere negation if it refers to something that cannot see such as a stone. But when not-seeing refers to a man, who in general can and should see, not seeing is considered a privation, blindness. Again some would try to consider such an absence a mere negation. But a blind person is a blind person; his entire being is modified by this absence. The reality of this blindness comes from the fact that this non-being exists in a certain kind of subject. It comes, therefore, from being. (Catalano 1985: 58)

Accordingly, Being and non-being are not separate but co-productive, mutually reinforcing forces that critically inform and shape each other.

The implications of this perspective for freedom are again immense. It avers that our being is shaped by a fundamental freedom of possibility, one that emerges from and catalysed by our incomplete relationship with Being. For Sartre (1956: 249), "Negativity as original transcendence is not determined in terms of a this; it causes a this to exist". Foundationally, it is a freedom to choose a future state of Being—an alternative existence—over a current essentialized self. Nothingness, in principle, is then a positive form of negation, the annihilation of what is for what could be. Consequently, "Totality can come to beings only by a being which has to be its own totality in their presence. This is precisely the case with the for-itself, a detotalized totality which temporalizes itself in a perpetual incompleteness" (Ibid.: 249).

Yet while Sartre speaks of despair and existentialism is by no means the most optimistic of philosophies, there still appears to be two rather key tensions in this account of the freedom of non-being. Notably, while the presence of lack may be universal, what it is and how it is understood can be quite subjective and relative. Surely, everyone feels as if they are missing something in their life, but how they experience this as a totality is certainly not uniform. In this respect, "A totality indeed supposes an internal relation of being between the terms of a quasi-multiplicity in the same way that a multiplicity supposes—in order to be this multiplicity—an inner totalizing relation among its elements" (Ibid.: 249).

As will be shown, the feelings of absence brought to the fore by capitalism generally and the free market specifically foster a wide array of totalistic interpretations of one's broader existential condition. Equally significant, is that people also live with a continual exposure to the possibility of total nothingness, the complete absence of Being—whether that be in terms of mortality or a dramatic fear that their very sense of self and place of the world is unassured—leading people to cling to their existence rather than risk the potential of annihilation.

The Fantasy of Freedom

For Sartre, human being is inexorably tied to non-being. The very existence of consciousness implies the negation of our current being through its questioning. Yet it also involves a fundamental fear of non-being—the consciousness that one could dissolve into nothingness. Tellingly Sartre distinguishes fear from anguish, the former denoting our terror at external forces inhibiting our freedom while the latter is the fraught realization of our own ability to freely extinguish our present being. He observes

> fear is fear of beings in the world whereas anguish is anguish before myself. Vertigo is anguish to the extent that I am afraid not of falling over the precipice, but of throwing myself over. A situation provokes fear if there is a possibility of my life being changed from without; my being provokes anguish to the extent that I distrust myself and my own reactions in that situation. (Sartre 1956: 65)

However, this rather stark though compelling distinction perhaps misses a fundamental fear, an almost primordial terror at the prospect of non-being. It is not merely that we are eternally "haunted" by nothingness but that we are therefore continually subjected to reminders of our own possible total negation.

Being then becomes an illusion to cling to, our psychic existence depending on the ontological security provided by our internalized social identity. It is understandable then that a central to our selfhood is, quoting famed social thinker Anthony Giddens (1991: 188), the attempt to "exclude, or reinterpret, potentially distributing knowledge … avoidance of dissonance forms part of the protective cocoon which helps maintain ontological security". This longing for ontological security also extends to our socially constructed group identities (Bion 1961; Johnson et al. 1970; Reicher and Hopkins 2001; Sherif et al. 1961). Thus while existence may proceed essence, it is ironically our essence that is perceived too necessary for us to exist at all.

Yet this also points to a deeper fundamental human existential longing. For Sartre, consciousness provides us with an inextinguishable sense of "lack", the desire to be more than what we currently are. Tellingly, he puts this in quite psychic terms, as an affective desire for transcendence. Quoting him at length:

> The existence of desire as a human fact is sufficient to prove that human reality is a lack. In fact how can we explain desire if we insist on viewing it as a psychic state; that is, as a being whose nature is to be what it is? A being which is what it is, to the degree that it is considered as being what it is, summons nothing to itself in order to complete itself. An incomplete circle does not call for completion unless it is surpassed by human transcendence. In itself it is complete and perfectly positive as an open curve. A psychic state which existed with the sufficiency of this curve could not possess in addition the slightest "appeal to" something else; it would be itself without any relation to what is not it. In order to constitute it as hunger or thirst, an external transcendence surpassing it toward the totality "satisfied hunger" would be necessary, just as the crescent moon is surpassed toward the full moon. (Sartre 1956: 87)

According to Sartre, people are haunted by this mythical ability to be the total creators of our reality like God. Importantly, he dismisses any notion of the divine as a "first cause", noting that they are human concepts through and through:

> This is equivalent to saying that being is uncreated. But we need not conclude that being creates itself, which would suppose that it is prior to itself. Being can not be causa sui in the manner of consciousness. Being is itself. This means that it is neither passivity nor activity. Both of these notions are human and designate human conduct or the instruments of human conduct. There is activity when a conscious being uses means with an end in view. (Ibid.: lxiv)

However, this desire for total agency, complete control over our existence, manifests itself into an almost primordial worship of Gods. The divine, in this respect, conforms to a vision of a completely free being, as

> every effort to conceive of the idea of a being which would be the foundation of its being results inevitably in forming that of a being which contingent as being-in-itself, would be the foundation of its own nothingness. The act of causation by which God is causa sui is a nihilating act like every recovery of the self by the self, to the same degree that the original relation of necessity is a return to self, a reflexivity; This original necessity in turn appears on the foundation of a contingent being, precisely that being which is in order to be the cause of itself. (Ibid.: 80–81)

Across the span of human history, the presence of the divine represents simultaneously the belief the absolute freedom is possible and the worldly lament that it continues to elude human grasp. Tragically, at least for Sartre,

> Every human reality is a passion in that it projects losing itself so as to found being and by the same stroke to constitute the In-it-self which escapes contingency by being its own foundation, the ens causa sui, which religions call God. Thus the passion of man is the reverse of that of Christ, for man loses himself as man in order that God may be born. But the idea of God is contradictory and we lose ourselves in vain. Man is a useless passion. (Ibid.: 615)

In place of the Gods is a scarier realization—that existence is not inherently meaningful. It is precisely at this moment that in conventional accounts of existentialism that our freedom fully makes itself known. However, it is also at this point that an even more fearful thought begins to creep in, namely that existence is uncontrollable, completely random, and that such pure freedom is therefore impossible—"a useless passion". As humans, it is surmised, we must stare into the existential abyss, forced to choose without clear or obvious choices.

This perspective echoes rather strangely given their stately different intellectual roots and inspirations, the insights of the French psychoanalyst Jacques Lacan. Drawing upon and ultimately dramatically reinterpreting

Freudian theories, Lacan argued that human development centred on the individual undergoing what he terms the "mirror stage" at a very young age. Our first encounter with our reflection brings to the fore both our embodied unity (I am this person) and our awareness that we are not fully physically in control of this self. Consequently, it reveals a "mirage" representing

> the total form of his body, by which the subject anticipates the maturation of his power in a mirage, is given to him only as a gestalt, that is, in an exteriority in which, to be sure, this form is more constitutive than constituted ... this gestalt, whose power [*prégnance*] should be considered linked to the species ... symbolises the *I*'s mental permanence, at the same time as it prefigures its alienating destination. This gestalt is also replete with the correspondences that unite the *I* with the statue onto which man projects himself, the phantoms that dominate him, and the automaton with which the world of his own making tends to achieve fruition in an ambiguous relation. (Lacan 2001: 95)

A further key part of our ego development is our association of this harmony between mind and body, intention and will, with a Big Other (in the first instance the father figure), a mythical state of wholeness that we desperately but futilely seek to regain as we mature. This split 'is operative in all of the various ways in which we fail to identify ourselves, grasp ourselves, or coincide with ourselves' (Bracher 2010: 113). Furthermore, our shared social and political identities are "always supported by a reference to a lost state of harmony, unity and fullness, a reference to a pre-symbolic Real which most political projects aspire to bring back" (Stavrakakis 1999). Tellingly, while Sartre is largely dismissive of traditional psychoanalysis for its "serious" ontological commitment to the unconscious, he analogously sees the Other as formative for identity as such:

> I am possessed by the Other; the Other's look fashions my body in its nakedness, causes it to be born, sculptures it, produces it as it is, sees it as I shall never see it. The Other holds a secret the secret of what I am. He makes me be and thereby he possess me, and this possession is nothing other than the consciousness of possessing me. I in the recognition of my object-state have proof· that he has this consciousness. By virtue of consciousness the Other is for me simultaneously the one who has stolen my

being from me and the one who causes "there to be" a being which is my being. Thus I have a comprehension of this ontological structure: I am responsible for my being-for-others, but I am not the foundation of it. It appears to me therefore in the form of a contingent given for which I am nevertheless responsible; the Other founds my being in so far as this being is in the form of the "there is." (Sartre 1956: 364)

It is this drive for psychic completeness that according to Lacan sustains us as beings. It allows humans to avoid the "real" of their existence which is fragmentary and unstructured. Accordingly, our existence—our "life"—is always intertwined with our potential non-being—our psychic "death" for Lacan as "the symbolic order is simultaneously non-being and insisting to be … it is a symbolic order in travail, in the process of coming, insisting on being realized' (Lacan 1988: 326). Hence,

[i]n our attempts to keep the ground from shifting, we idealize our objects, clothes, people and thoughts with the ideological garments of the good, the true and the beautiful. But the unconscious moves as desire, tracking evacuated jouissance, as desire and sexuality dancing in the endless vacillation around a void. (Ragland 2013: 100)

Rather, being revolves around the construction of an illusionary "stable" reality, one that marked by its coherence and the possibility of attaining wholeness.

These ideas resonate with and to an extent add to the existentialism put forward by Sartre. Freedom exists both as an illusionary but sustaining means and end for humans. It is the ultimate state they strive for and the principle means through which they pursue this eternally out of reach utopian aspiration. Lacan refers to this as a fantasy, an affective discourse, accordingly described by the renowned Lacanian commentator Slavoj Žižek (1993: 201) as an affective discourse

which holds together a given community [that] cannot be reduced to the point of symbolic identification … the bonds linking together its members always implies a shared relationship to the Thing, toward enjoyment incarnated … If we are asked how we can recognise the presence of this Thing, the only consistent answer is that the Thing is present in that elusive entity called our "way of life".

In this respect, we are psychically "gripped" by a comprehensive fantasy of freedom. Consciousness, in turn, is shaped by this unconscious drive for wholeness and illusionary freedom. While these fantasies are individually experienced and internalized, they are culturally proscribed. Accordingly, Sartre's original idea was that it is in consciousness that we are made acutely and concretley aware of our existential gap. Yet it is an absence of Being that is inexorably linked to a cultural fantasy of freedom that is not of our choosing and from which we seemingly cannot psychically escape.

Desiring Freedom

Theoretically freedom is often understood as a substantive set of either intrinsic or social values or practices. However, within the popular lexicon, it is commonly conceived as an emotive longing, a psychological desire to resist tyranny and be free from unfair social constraints. It is a romantic wish as much as it is a comprehensive or completely rationally considered political agenda. It is crucial, therefore, to understand how and in what way people desire freedom.

The reframing of freedom as a cultural fantasy is absolutely critical for such an exploration. The fundamental yearning to be free is channeled into particularized social discourses of freedom. Thus market freedom stands as the dominant ideal for realizing this ultimate longing to be emancipated from economic and social burdens while having the agency to negotiate, escape, and even subvert prevailing power relations. (Bloom and Cederstrom 2009; Bloom 2016). To this effect, freedom becomes fetishized, evolving from a universal drive to be free to a particular and ideologically narrow desire to be free in a certain way. Returning again to the insights of Žižek (2004: 94),

> The ontological scandal of the notion of fantasy resides in the fact that it subverts the standard opposition of "subjective" and "objective". Of course, fantasy is, by definition, not "objective" (in the naïve sense of "existing independently of the subject's perception"). However, it is also not "subjective" (in the sense of being reducible to the subject's consciously experienced intuitions). Rather, fantasy belongs to the "bizarre category of the objectively subjective—the way things actually objectively seem to you even if they don't seem that way to you".

Market freedom, hence, is put forward and internalized as the sole provider of individual and collective agency.

This freedom fetish stands as a tempting but also elusive ideal social aspiration. It exists, to use a Lacanian term, as an object petit (Lacan 2001). People continually invest their hopes in this hegemonic freedom, having faith that it will someday allow them to realize their dreams. Freedom is experienced, accordingly, both as an everyday reality as well as an eternally desirable but always just out of reach desire. This reflects the simultaneous and mutually reinforcing "beatific" and "destabilizing" aspects of fantasy, as

> On the one hand, fantasy has a beatific side, a stabilizing dimension, which is governed by the dream of a state without disturbances, out of reach of human depravity. On the other hand, fantasy has a destabilizing dimension, whose elementary form is envy. It encompasses all that "irritates" me about the Other, images that haunt me about what he or she is doing when out of my sight, about how he or she deceives me and plots against me, about how he or she ignores me and indulges in an enjoyment that is intensive beyond my capacity of representations, etc. (Žižek 1998: 192)

Encountered then is another profound paradox of freedom—it is in its relative absence that it is at its personally and socially strongest. It is precisely its non-being that makes its potential Being so tantalizing.

The contradictory desire for a freedom that is eternally on the horizon plays a fundamental role, ironically, in sustaining people as social subjects. These ongoing efforts to achieve and embody this socially proscribed freedom offer ontological security, a clear and stable framework for ordering and acting within a relatively constant cultural reality. The pursuit of freedom is consequently necessarily a "failed identification" (Stavrakakis 1999). And it is from this failure that paradoxically identity is temporarily assured and enjoyment is obtained. There is what Lacan refers to as a deep jouissance derived from pursuit of a fetishized freedom, manifested in its daily incomplete realization and in the possibility of its prospective attainment.

Beyond the individual, social stability is maintained through keeping open this dominant "gap of freedom". Here the existential threat of non-being (as well as its potential excitement) is displaced by a secure quest for a particular freedom. It is a "safe" form of nothingness, a manageable absence that attaches one to a specific form of Being. This echoes the ways in which resistances can form a "safe" symbiotic relationship with the very power in which it is ostensibly opposing (Bloom 2013). Even more fundamentally, the very dynamic of power and resistance, whereby a dominant order must be challenged, creates is own imprisoning form of "safety". Paradoxically, every resistance, in this regard, is

> inherently to a degree 'safe'. Safety here is understood not principally as 'making safe' the position of 'power holders'; rather, it refers to the onto-logical security resistance grants individuals. This 'safety', in turn, deepens our identification with prevailing power relations, our identity secured in our production as resistance subjects. (Ibid.: 232)

Similarly, the constant efforts to achieve freedom, to fight off the fear of others and the anguish at our own selves, become an "ontologically secure" mode of being. The different content of these struggles masking a shared and quite analogous framework for guiding our thought and action. Hence all attempts to be free are shadowed always by a certain degree of "safety".

Ontological security is as such inexorably linked to as Sartre first gestured to nothingness. Yet it is a nothingness that is tamed and singularly associated with a particularized version of existence—one revolving around the eternal disappointment with and hopeful pursuit of an idealized fantasy of Being. However, it also reflects, for Sartre, the internalization of our bad faith, the surrendering of our freedom to our perceived unconscious. In his words,

> Thus psychoanalysis substitutes for the notion of bad faith, the idea of a lie without a liar; it allows me to understand how it is possible for me to be lied to without lying to myself since it places me in the same relation to myself that the Other is in respect to me; it replaces the duality of the deceiver and the deceived, the essential condition of the lie, by that of the "id" and the "ego". (Sartre 1956: 51)

Consequently, the appeal of market freedom is in part as a cultural aspiration, one that elides its daily experiences of regulation and coercion for a romanticized account of using the free market to control one's own destiny. It is also found in our unconscious investment in it, the belief that it is merely a reflection of our "human nature" and as such a stable foundation for us to obtain ontological security.

The "reality" of freedom is then found in its constant unfulfilled desire. The free subject is the socially constructed and psychically imprisoned desiring subject. Their very survival as a self is inexorably linked with this futile but ironically enjoyable pursuit of these culturally entrenched freedom fantasies. The more fundamental presence of existential freedom, the ability to truly and genuinely shape your given historical circumstances, is articulated and unconsciously experienced as a fatal encounter with our own non-being. Non-being is transformed from a demand for the scope of Being to be expanded to a fear that without this narrow reality we would disintegrate into an inescapable nothingness. It is exactly this underlying fear, this perceived utter precariousness of existence as such, that currently consigns us to the tightening psychic grip of capitalist being.

Capitalist Being and Nothingness

The free market is most often portrayed as being driven by rationality, not emotion. At its most pure, it envisions completely rational actors attempting to maximize their utility. The actuality of capitalist existence is, of course, far more emotionally complex and psychologically driven. It relies on a rich array of affective desires for its reproduction. The free market in particular nurtures and preys upon deep-seated longings for professional success, personal well-being, social agency, and increasingly ethical fulfilment.

At the core of market freedom is a profound experience of lack. In particular, it revolves around personal feelings of incompleteness. To this end, renowned psychoanalytic social theorist Todd McGowan links capitalism to an eternally unfulfilled desires for "belonging":

> The capitalist subject constantly experiences its failure to belong which is why the recurring fantasy within capitalism is that of attaining some degree of authentic belonging (in a romantic relationship, in a group of friends, in

the nation, and so on). Though capitalism spawns this fantasy it constantly mitigates against this fantasy's realization. Capitalism offers the promise of belonging with every commodity and with the commodity as such but the subject can never buy the perfect commodity, or enough of them to unlock the secret of belonging. (McGowan 2016: 20)

This is readily apparent in its hyper-consumer culture. New products promise to fulfil us, make us whole, ensure that we are more attractive, stylish, and intelligent. It even helps to ease the pressure of buying your way to an ideal self. Indeed, the modern marketplace is full of products—whether it be unhealthy food or entertainment items—that provide people temporary comfort from the drudgery of their daily lives and the demands to achieve perfection through consumption.

At a deeper level, this culture of fulfilment by consumption reflects the psychic enjoyment obtained from this capitalist lack. Put differently, there is an existential safety, a clear and stable sensible reality, in the ongoing effort to purchase happiness and health. It is in the non-being, the potentiality to become that which we are currently not, that our capitalist selves are most assured. In this respect, it is in the nothingness—or more precisely the not yet something—where the free market persists and thrives. Ironically, it is by feeding into our deepest insecurities that capitalism most guarantees our ontological and subjective security. Paradoxically, the more psychologically and emotionally insecure we are, the more fundamentally secure (and ultimately trapped) we remain as capitalist subjects.

The association of capitalism—and the free market—with the constant pursuit of pleasure tells only part of its story. Underlying this surface level excess is an underlying enjoyment in the lack of pleasure it provides. The feeling of needing more, of never being satiated whether professionally or as a consumer, is a perennial reminder that one exists and will persist, that even if they remain unfulfilled today the ultimate wholeness they seek is still possible. There is therefore an exquisite enjoyment to be personally gained in our own exploitation, an exchange of existential freedom for the safer market freedom. According to the critical theorist Ciara Cremin (2010: 131), "the subordination to capital (the material fact of labour) is defused by the sense we have of our independence from the employer (an identification that is not associated with the act of labour)". It is, therefore, in the appearance of nothingness that capitalist Being is assured and perverse enjoyment gained from this perpetually dissatisfied self is guaranteed.

This psychoanalytic perspective radically reconsiders Marx's famous canard about the insatiable nature of capitalism. Capitalist and capitalism, according to Marx, is driven by an unquenchable drive for profit and new markets. In this sense, he proclaims that "Capital is dead labour, which, vampire-like, lives only by sucking living labour, and lives the more, the more labour it sucks" (Marx 1976: 342). He likewise writes that "the vampire will not let go" as the extension of the working day "only slightly quenches the vampire thirst for the living blood of labour" (Marx 1976: 367). No amount of money will ever fully satiate this need; no new successful business venture will make this longing for greater economic gain to go away. There are always fresh business opportunities to exploit, novel ways to make one's business more efficient and profitable.

Even if individually some manage to escape this passionate craving to be ever richer, structurally the entire premise of capitalism relies on this perpetual need for greater profit. Economies must be constantly growing or risk stagnation and death. There is no end point, only expansion—as the promise of the market is never a utopian future but only a receding horizon promising the possibility of discovering new opportunities for profit. Psychically, it is this constant lack that sustains as beings and in doing so maintains Capitalist Being. If it was ever filled, if the desire ever vanished, then its whole reality (let alone its economy) would come crashing down. Jones and Spicer (2005: 237) note,

> [o]ne secures identity not in "being" an enterprising subject but in the gap between the subject and the object of desire. Not only does it not matter that the object is unattainable. This lack is central to maintaining desiring. And, as Lacan indicates, if we ever achieve the object of desire, it collapses—it falls apart and is changed inexplicably into a gift of shit.

To avoid this seemingly apocalyptic fate, people invest in their own personal and shared affective capitalist history. Lacan differentiates history proper from what he describes as a "psychic history":

This limit is present at every instant in what is finished in this history. It represents the past in its real form; it is not the physical past whose existence is abolished, nor the epic past as it has become perfected in the work of memory, nor the historical past in which man finds the guarantor of his future but rather the past which manifests itself in an inverted form in repetition. (Lacan 2001: 262)

It is in the love and strife that death is temporally kept at bay. Specifically, by keeping open a "safe" capitalist gap of non-being, the more fundamental and fearful threat of total annihilation is put off. We thus enter into an enduring, though always precarious, affective capitalist history in which our current non-being becomes an as yet unrealized dream of what we may one day become. What is crucial is not what we actually become but rather that at all times we still have the opportunity to become something.

In sum, our very being seemingly depends on us existing as infinitely desiring capitalist subjects. There is nothing beyond capitalism or the free market. The only existential solace we have is in the lack provided to us by insatiable market. Non-being is confined to either a fundamental threat of total psychic disintegration or capitalist approved consumer and professional aspirations. In this respect, the sky truly is the limit of our dreams, contained always by an impenetrable horizon of capitalism. The existential excitement of an existence that transcends our present Being is replaced by a resignation that the only reality we could ever "realistically" inhabit is one eternally caught between the poles of Capitalist Being and nothingness.

Enjoying Existential Freedom

Both capitalism and existential freedom prioritize the importance of choice. At stake is the ability of individuals and communities to choose their destiny, to not be defined by their historical circumstances. For the market this is translated into being able to "freely" choose where one works, what one buys, and if successful how one would like to live. Yet ultimately, these are simply market options—an experience of freedom curtailed to exclusively reflect capitalist prerogatives at all times. The "real" choice it provides people is the false existential dilemma between an incomplete Capitalist Being and complete and total nothingness. It brooks for no type of existence, no manifestation of Being, beyond its limited ideological and exploitive horizons. The only "reality" that is assured and possible is one defined by the free market.

Significantly, it transforms the consciousness of non-being, that recognized absence of what we may still yet become individually and collectively, from a systematic to a personal lack. To this effect

> "The present moment is marked by anxieties about society falling apart, and nostalgia for a lost era of social cohesion," and [t]hese anxieties shape the dominant narrative about the causes of the recession—which are seen as resulting not from the excesses of the financial sector but from a profligate welfare system and an overly permissive immigration system, which has given the wrong people access to public services—the unemployed, the disabled, single parents and immigrants. (Forkert 2014: 41)

More precisely, it is not that the free market is lacking in any substantial way—or in any way that can be supposedly realistically changed. Rather, it is that we as individuals or communities are lacking and need to strive for fulfilment. And it is exactly in the perverse psychic enjoyment of this lack that capitalism is continually materially and socially reproduced.

Capitalism is therefore maintained as a cultural fantasy that offers subjects a precarious sense of ontological security and psychic stability. Importantly, it is experienced less as an illusion (and indeed any envisioning of a capitalist alternatively is ironically dismissed as mere fantasy) and more as an "objective reality". Further, these fantasies are by no means singular or static. They represent a universal "Law" upon which different fantasies can emerge and flourish. For Lacan (2001: 66),

> This law, then is revealed clearly enough as identical with an order of language. For without kinship nominations, no power is capable of instituting the order of preferences and taboos that bind and weave the yarn of lineage through succeeding generations'.

To this effect, while everyone must universally adhere to the Law of Capitalism, they invest in quite personalized fantasies of market fulfilment.

There is a need, therefore, to go beyond these capitalist fantasies. From an almost purely psychoanalytic perspective, Lacan proposes the potential to "traverse" or transcend these psychic fantasies. Hence,

> What psychoanalysis can do to help the critique of ideology is precisely to clarify the status of this paradoxical jouissance as the payment the exploited, the served received for serving the master. This jouissance of course, always emerges within a certain phantasmic field; the crucial precondition for breaking the chains of servitude is thus to "transverse the fantasy" which structures our jouissance in a way which keeps us attached to the Master—makes us accept the framework of the social relationship of domination. (Žižek 1997: 48)

At the social and organizational level, this has been developed into a radical politics of transcending our cultural fantasies (see Hoedemaekers 2009). Particularly relevant to questions of freedom is the crucial need for people to traverse existent fantasies of freedom such as those associated with the market. Put differently, to refuse to accept that the only way choices we have for shaping our existence are the ones provided by a dominant discourse like that of capitalism.

Such a radical break with prevailing forms of freedom reflects the embrace of an even deeper existential choice. It is to reject the false binary of an existing in conformity to current manifestation of Being or having to face total subjective annihilation . To do so requires much more than just a rational commitment to the unknown, a positive affirmation of the exciting potential instead of fearful threat of non-being. Rather, it demands a reconfiguration of psychic enjoyment to reflect this investment in the possibilities of nothingness. Glynos (2000), in this regard, contrasts an "ethics of desire" from an "ethics of the drive" focusing on the enjoyment gained by always traversing fantasies and experiencing our fundamental lack. Mapped onto a theory of existential freedom, there is a profound jouissance obtained from this constant drive to go beyond the limits of Being, the freedom to explore the potentialities of an incomplete existence.

At stake is nothing less than the personal and mass ending of capitalist history. As discussed in the above section, history at least as it is affectively experienced is not defined by the passing of events. Conversely, it is the perpetual cycle of love, strife, and death that is repeated ever more in relation to a culturally proscribed desire. This echoes, Sartre's own objections to the ways psychoanalysis represses existential freedom through subscribing all our actions to a predetermined developmental past:

> Consequently the dimension of the future does not exist for psychoanalysis. Human reality loses one of its ekstases and must be interpreted solely by a regression toward the past from the standpoint of the present. At the same time the fundamental structures of the subject, which are signified by its acts, are not so signified for him but for an objective witness who uses discursive methods to make these meanings explicit. No pre-ontological comprehension of the meaning of his acts is granted to the subject. And this is just, since in spite of everything his acts are only a result of the past, which is on principle out of reach, instead of seeking to inscribe their goal in the future. (Sartre 1956: 458)

It is crucial, therefore, to "drive" past this history, deconstructing it and breaking free from its psychic grip over our existence.

The immediate objection that can be posed is that such a revolutionary remaking of history, this sweeping departure from capitalism, is not purely a matter of mentally transcending our psychic imprisonments. There are real material, social, and political chains that also need breaking. However, this radical enjoyment is by no means separate from political struggle. At play is the forming of identity around a revolutionary lack, the desire to experience an alternative to that which presently exists and dominates our social imagination. According to renowned political theorist Jodi Dean (2016), collective movements open up "gaps of possibility" for reshaping economic and cultural relations, one manifested in the transition from impassioned crowds to revolutionary Parties. This freedom experienced in and through a Party as described by Dean is linked to concrete practices and relations, the affective embrace of fresh ways of Being.

An absolutely fundamental philosophical and political question, then, is how do we enjoy freedom? Or put differently, what type of freedom do we invest our sense of selfhood in—the fantasy of agency provided to us or the drive to go beyond that Being currently given to us? Existential enjoyment is the jouissance discovered in an eternal and exciting relationship with nothingness, a view of non-being as an opportunity for experimenting with new ways of Being as opposed to a sentence to a permanent subjective death. It is the view of social "facts" as pleasurable "facticities" for remaking existence, the transformation of a demanding existential freedom into an enjoyable one. It is precisely this revolutionary enjoyment of Being and nothingness that fully permits us to become free existential subjects.

References

Bauman, Z. (2006). *Liquid Times: Living in An Age of Uncertainty.* Cambridge: Polity Press.

Bion, W. (1961). *Experiences in Groups and Other Papers.* London: Tavistock.

Bloom, P. (2013). The Power of Safe Resistance. *Journal of Political Power, 6*(2), 219–239. https://doi.org/10.1080/2158379x.2013.805919.

Bloom, P. (2014). Back to the Capitalist Future: Fantasy and the Paradox of Crisis. *Culture and Organization, 22*(2), 158–177. https://doi.org/10.1080/1 4759551.2014.897347.

Bloom, P. (2015). Work as the Contemporary Limit of Life: Capitalism, the Death Drive, and the Lethal Fantasy of 'Work–Life Balance'. *Organization, 23*(4), 588–606. https://doi.org/10.1177/1350508415596604.

Bloom, P. (2016). *Beyond Power and Resistance: Politics at the Radical Limits.* London: Rowman & Littlefield.

Bloom, P., & Cederstrom, C. (2009). "The Sky's the Limit": Fantasy in the Age of Market Rationality. *Journal of Organizational Change Management, 22*(2), 159–180. https://doi.org/10.1108/09534810910947190.

Bracher, M. (2010). *Lacanian Theory of Discourse.* New York: New York University Press.

Catalano, J. (1985). *A Commentary on Jean-Paul Sartre's Being and Nothingness.* Chicago: University of Chicago Press.

Clarke, J. (2014). After Neoliberalism? Markets, States, and the Reinvention of Welfare. In M. Hayward (Ed.), *Cultural Studies and Finance Capitalism: The Economic Crisis and After* (pp. 103–112). London: Routledge.

Cremin, C. (2010). Never Employable Enough: The (Im)possibility of Satisfying the Boss's Desire. *Organization, 17*(2), 131–149. https://doi.org/10.1177/1350508409341112.

Dean, J. (2016). *Crowds and Party*. London: VERSO.

Forkert, K. (2014). The New Moralism: Austerity, Silencing and Debt Morality. *Soundings, 56*(56), 41–53. https://doi.org/10.3898/136266214811788808.

Gabriel, G. (2009). A Neo-Liberal Nihilism? *Open Democracy.*

Giddens, A. (1991). *Modernity and Self-Identity – Self and Society in the Late Modern Age*. Cambridge: Polity Press.

Giroux, H. (2011). *Zombie Politics and Culture in the Age of Casino Capitalism*. New York: Peter Lang.

Glynos, J. (2000). Thinking the Ethics of the Political in the Context of a Postfoundational World: From an Ethics of Desire to an Ethics of the Drive. *Theory and Event, 4*(4). ISSN:1092-311X.

Harman, C. (2009). *Zombie Capitalism: Global Crisis and the Relevance of Marx*. New York: Guilford Publication.

Hayward, M. (2010). The Economic Crisis and After. *Cultural Studies, 24*(3), 283–294. https://doi.org/10.1080/09502381003750260.

Hoedemaekers, C. (2009). Traversing the Empty Promise: Management, Subjectivity and the Other's Desire. *Journal of Organizational Change Management, 22*(2), 181–201. https://doi.org/10.1108/09534810910947208.

Johnson, N., Middleton, M., & Tajfel, H. (1970). The Relationship Between Children's Preferences for and Knowledge About Other Nations. *British Journal of Social and Clinical Psychology, 9*(3), 232–240. https://doi.org/10.1111/j.2044-8260.1970.tb00669.x.

Jones, C., & Spicer, A. (2005). The Sublime Object of Entrepreneurship. *Organization, 12*(2), 223–246. https://doi.org/10.1177/1350508405051189.

Lacan, J. (1988). *The Seminar of Jacques Lacan. Book II: The Ego in Freud's Theory and in the Technique of Psychoanalysis, 1954–1955*. New York: W.W. Norton and Company.

Lacan, J. (2001). *Ecrits*. New York: W.W. Norton and Company.

Lanci, Y. (2014). Zombie 2.0: Subjectivation in Times of Apocalypse. *Journal for Cultural and Religious Theory, 13*(2), 25–37.

Marx, K. (1976). *Capital: A Critique of Political Economy* (Vol. I). London: Harmondsworth.

McGowan, T. (2016). *Capitalism and Desire: The Psychic Costs of Free Markets*. New York: Columbia University Press.

Ragland, E. (2013). *Essays on the Pleasures of Death.* Hoboken: Taylor and Francis.

Reicher, S., & Hopkins, N. (2001). Psychology and the End of History: A Critique and a Proposal for the Psychology of Social Categorization. *Political Psychology, 22*(2), 383–407. https://doi.org/10.1111/0162-895x.00246.

Sandel, M. (2016). Bernie Sanders and Donald Trump Look Like Saviours to Voters Who Feel Left Out of the American Dream. *The Guardian.*

Sartre, J. (1956). *Being and Nothingness.* New York: Gallimard.

Seib, G. (2016). Behind the Rise of Populism, Economic Angst. *The Wall Street Journal.*

Sherif, M., Harvey, O., Hood, W., Sherif, C., & White, J. (1961). *The Robbers Cave Experiment.* Middletown: Wesleyan University Press.

Shiller, R. (2017). The Economy Is Stagnant Because People Fear for the Future. *The Guardian.*

Stavrakakis, Y. (1999). *Lacan and the Political: Thinking the Political.* New York: Routledge.

Žižek, S. (1993). *Tarrying with the Negative.* London: Verso.

Žižek, S. (1997). *The Plague of Fantasies.* London: Verso.

Žižek, S. (1998). The Seven Veils of Fantasy. In D. Nobus (Ed.), *Key Concepts of Lacanian Psychoanalysis.* London: Rebus Press.

Žižek, S. (2003). *Tarrying with the Negative.* Durham, NC: Duke University Press Books.

Žižek, S. (2004). What Can Psychoanalysis Tell Us About Cyberspace? *The Psychoanalytic Review, 91*(6), 801–830. https://doi.org/10.1521/prev.91.6.801.55957.

6

Subjected to the Free Market: The Subject of Existential Freedom

The free market is largely considered to be simply a matter of human nature. Individuals are naturally self-interested and will therefore just as naturally seek to maximize their utility. The market is put forward as a timeless part of the human condition, a hallmark of all societies spanning our historical existence. The idea of a free market, hence, is seemingly redundant. It is merely a reflection of who we are at our most essential and ordinary. Further, it is simply fundamental to being free at all. Quoting one of its most famous—or depending on perspective infamous—proponents, economist Milton Friedman (2009: 9),

> Historical evidence speaks with a single voice on the relation between political freedom and a free market. I know of no example in time or place of a society that has been marked by a large measure of political freedom that has not also used something comparable to a free market to organize the bulk of economic activity.

In practice, there requires quite a bit of social engineering to produce the market subject as highlighted in Chap. 4. People need to be taught to be entrepreneurial, to be a good employee, to manage their finances, and so on. In this respect, neoliberalism should not be confused with a

© The Author(s) 2018
P. Bloom, *The Bad Faith in the Free Market*,
https://doi.org/10.1007/978-3-319-76502-0_6

complete rejection of government or a total embrace of libertarianism. If "at the one end of the line is 'anarcholiberalism,' arguing for a complete laissez-faire, and the abolishment of all government. At the other end is 'classical liberalism,' demanding a government with functions exceeding those of the so-called nightwatchman state" (Blomgren 1997: 224, quoted in Thorsen 2010), then neoliberalism exists in the middle, championing an active state whose purpose is to protect and spread the free market.

Indeed the role of the state under neoliberalism is not so much in retreat but reconfigured and in some ways expanded. The government is meant to be educative, to foster a free market society and produce more market-oriented and marketable subjects (Levi-Faur and Jordana 2005). Consequently, while

> neoliberalism is often understood as a synonym of the doctrine of self-regulating markets ... neoliberals are quite clear that states have certain functions to fulfill in order to make markets function. But they should only engage in certain kinds of actions, and these are particularly market-enabling actions. (Biebricher 2015: 1)

The last chapter explored the free market as an invasive cultural fantasy of freedom upon which the self is formed. In this spirit, freedom can also be profoundly socially disciplining, regulating individual actions in conformity with dominant values and expectations. As famed cultural anthropologist Ortner (1995: 86) presciently notes, "every culture, every subculture, every historic moment, constructs its own forms of agency, its own modes of enacting the process of reflecting on the self and the world and of acting simultaneously within and upon what one finds there." In the case of market freedoms, the demand to maximize one's utility, to be economically "rational", to find ever new ways to make a profit is extremely inscriptive.

To this effect, people are expected to become living, breathing, thinking objects of capital. Put differently, they must creatively determine how to optimize their market value and max out their happiness as consumers. Whereas capitalist oppression is commonly linked merely to the shaping of our "rational" desires, in the modern age there is also a more disturbing possibility: that the critique of individualism and monetary calculation is now being incorporated into the armoury of utilitarian policy and management. One of the key insights of behavioural economics is that, if one wants to control other human beings, it is often far more effective to appeal to their sense of morality and social identity than to their self-interest. (Davies 2015)

They are thus free to be more efficient, more productive, and more economically useful as well as more ethical and better "good citizens" (see Bloom 2017). They become a social being to be internally and externally moulded to fit these economic demands. Returning again to the insights of Davies (2014: 13)

for social relations to be organised into reasonably persistent, reliable and peaceful institutions, at some point there must be a shared sense of normativity, a shared basis on which to distinguish between people and between things and make evaluations of their relative worth.

Thus, freedom is premised ironically on fulfilling these social pressures, their choices confined to these requirements that they become more self-regulating.

Consequently, people are increasingly being subjected to the free market. In this respect, freedom turns into that which is repressive, a social force that people are resigned to accept rather than something that is liberating. Market freedom is a burden, dictating, influencing, and in part determining our actions and conceptions of what is possible and why. Our existence and potential as subjects is an outgrowth of our being subjected to and ultimately dominated by market freedom. It is our embodied knowledge of what freedom is, how we can and should manifest as an agency that shapes who we are as acting and thinking people.

Yet what must not be forgotten is that just as every society produces its own knowledge of power, so does every relation, institution, organization reflect a specific mode of freedom. As such the knowledge/power binary first theorized by Foucault is double-edged. At its sharpest, it is a timely and perpetual reminder that all our social interactions contain within them the seeds that grow into the power relations that direct our lives.

Quoting the famous French social thinker Foucault—whose ideas will be drawn upon throughout the remainder of this chapter—"Where there is power there is resistance, and yet, or rather consequently, this resistance is never in a position of exteriority to power" (1978: 95–96). The same can be said about freedom, power produces freedom and yet is also always confronted and challenged by it.

Nevertheless, it is also empowering, as it grants us the power to meet these cultural expectations—it is the "how" of power, how we realize and continually enact our social reality and the power relations within it. The radical edge of this empowerment is that it opens the space for people to experience a more fundamental existential freedom, the material and social means to shape their environment and their place within it. He promotes, therefore, the "space of freedom understood as a space of concrete freedom, i.e. of possible transformation" (Foucault 1990: 36). Such freedom provides them with the knowledge and ability to experiment with their own selves and inhabit alternative ways of Being. In transforming them into a subject of power, it also holds open the possibility for them to also become a subject of freedom.

Free Market Subjects

The free market is heralded in almost utopian tones as the ultimate harbinger of individual liberty. It contrasts itself with systems that promote greater public intervention as well as those that culturally prioritize collectivism over individualism. These sentiments are epitomized by the twentieth-century libertarian thinker Ayn Rand—whose "objectivist" philosophy has made her a hero among those free market Right. She proclaims "Money is only a tool. It will take you wherever you wish, but it will not replace you as the driver." Similarly, she rails against the tyranny of the state:

> We are fast approaching the stage of the ultimate inversion: the stage where the government is free to do anything it pleases, while the citizens may act only by permission; which is the stage of the darkest periods of human history, the stage of rule by brute force.

However, equally important is the effective social construction and promotion of a successful free market subject. Entrepreneurs are not simply born, they must be taught and culturally nourished (see Fiet 2001). This represents a marked different than the traditional notion of the capitalist subject. The "ideal employee" of old is being rapidly replaced by the high-flying entrepreneur. What is crucial is that embodying this free market ideal is seen to be equivalent to being free itself (Kreft and Sobel 2005). As one recent *Forbes Magazine* article proclaimed "Entrepreneurship Is Ticket To Freedom" since "if your deepest yearning is to be free to lead your work life the way you want, regardless of what reponsibilities [sic] come with it, entrepreneurship is worth considering. It apparently delivers what it promises on that front" (Pofeldt 2014). Previous ideas of political representation or industrial democracy are now subsumed into an almost all-encompassing vision of the free market subject. Tellingly,

> Free markets were supposed to lead to free societies. Instead, today's supercharged global economy is eroding the power of the people in democracies around the globe. Welcome to a world where the bottom line trumps the common good and government takes a back seat to big business. (Reich 2009)

This sense of personal market liberation is wide-ranging—extending from the popularized financial "master of the universe" to the ethically committed "social entrepreneur".

The idea of an influential and all-powerful free market subject sits awkwardly with the increasingly perceived depersonalized nature of modern capitalism. Emerging technologies associated with Big Data, analytics, artificial intelligence, and automation have led, if anything, to the growing sense of dehumanization. Introducing the threat of what she terms "Weapons of Math Destruction", writer Cathy O'Neil (2017) notes "Ill-conceived mathematical models now micromanage the economy, from advertising to prisons ... there opaque, unquestioned, and unaccountable and they operate at a scale to sort, target, or "optimize" millions of people". This reimagines builds on the famous image of the little tramp factory worker lost amidst the production line clogs in Chaplin's classic film *Modern Times*. The updated Post-Modern Times is one where people lost in the face of the algorithms that control their personal lives and professional fate.

These mounting feelings of depersonalization create resurgent desires for an empowered sovereign. To this effect,

> the identification with a powerful sovereign provides individuals with ontological security in the face of rather complex micro-processes of power and broader depersonalised forms of subjection associated with neoliberalism. In this respect, individuals are affectively 'gripped' by sovereignty to account for the complexity and incoherence associated with the concrete and discursive operation of disciplinary power. The appeal of a sovereign fantasy lies in its promise of granting individuals a sense of 'sovereign' agency perceived to be lacking in their existence as 'agency-less' disciplinary subjects of neoliberalism. (Bloom 2015: 9–10)

These desires are manifested in the longing for competent and ethical leadership (Rhodes and Bloom 2012). More recently, this has played out in the appeal of the renegade populist leader and anti-hero for whom the standard rules do not apply (Tie 2004).

These represent renewed longings for experiencing a sense of existential freedom. At their heart, whether they be good leaders or perversely beloved anti-heroes, is a strong need to feel once again in control of their destiny. To rail against the encroachment of a social system that feels "inevitable" and unchangeable by human individuals or communities. The backlash against such inevitability, the populism from both the left and the right, indicates the wish to place our fate back into human control.

The prospect of such freedom is increasingly promised to the powerful individual who can rail against technological and social constraints. Interestingly, this has given rise to the romanticizing of the corporate executive both economically and politically. Specifically

> The changes to leadership heralded by neoliberalism are witnessed in the evolving image of it adopted by elected politicians and by corporate managers. The ideal politician is increasingly one who embodies business values of "efficiency" and "profitability". Referred to popularly at times as the "CEO President", this emerging model of political authority reflects the government's perceived role in maximizing their country's economic competitiveness within a volatile global market. (Bloom and Rhodes 2016: 360)

Such neoliberal processes of "executization" expose the complete conflation of freedom with corporate power. The goal is not the creation of social alternatives but rather competing against others and ultimately ruling over them (Bloom and Rhodes 2018). It also contains within it the glint of existential freedom—as it promotes the visionary qualities of such business titans.

Yet this also suggests a fundamental and rather insidious free market caveat to being able to experience this deeper freedom. There is a fine print condition on being able to be existentially free, in this respect. Notably, you can pursue any type of life you want, strive to achieve any goals, be an utter visionary and reshape the future, just so long as you behave and embody a corporate way of Being. Here the free market is not so much an end in itself but the exclusive means, the only knowledge available, to personally and collectively emancipate ourselves from our present existence for something profoundly and radically different. The current strive for total freedom we are ironically continually produced and reproduced always as free market subjects.

Objectification and Existentialism

A major tension plaguing human existence is whether we as people are in fact free subjects or merely social objects. From philosophy to psychology to sociology to neurology, the question of our free will, or lack thereof, remains paramount. Such questions are perhaps especially important for any attempts to understand and enact existential freedom. Indeed, the entire premise of existentialism is that individuals—and by association arguably societies—do not only have free will but that they are fated to constantly put into action through our choices. It is, in this respect, both our birthright and our curse.

This perspective runs directly counter to other radical traditions—particularly those associated with Marxism—that focus instead on processes of objectification. Here, the emphasis is on the transformation of individuals into social objects, living and breathing reproductions of economic demands and desires. Specifically, it is noted that people under capitalism are turned into mere objects of capital, their bodies and minds shaped and directed to reflect the needs of capital (Lukacs 1971). These chime with, though are not identical to, post-structural account that in different ways stress how social norms dictate personal beliefs, actions, and even selfhood.

Sartre has his own theory of objectification, albeit one that is informed by his broader existentialist philosophy. He notes that people can reject their inherent freedom to become merely a social object becoming in effect a "me-as-object". Consequently

> I am my own detachment, I am my own nothingness; simply because I am my own mediator between Me and Me, all objectivity disappears. I can not be this nothingness which separates me ·from me-as-object, for there must of necessity be a presentation to me of· the object which I am. Thus I can not confer on myself any quality without mediation or an objectifying power which is not my own power and which I can neither pretend nor forge. (Sartre 1956: 274)

As such, they are not freely choosing their existence but rather living up and seeking constantly to emulate to an existent cultural ideal. In this respect,

> my being-for-others—i.e., my Me-as-object—is not an image cut off from me and growing in a strange consciousness. It is a perfectly real being, my being as the condition of my selfness confronting the Other and of the Other's selfness confronting me. It is my being-outside-not a being passively submitted to which would itself have come to me from outside, but an outside assumed and recognized as my outside. (Ibid.: 285–286)

As such, one's entire life is wrapped up in their daily choice to conform to prevailing social expectation, to embody in both their thought and actions an identity that is not their own.

It is precisely this choice to be objectified that Sartre refers to as bad faith. Notably, it describes a state where people have turned their back on their freedom, rejecting the ability to explore different ways of being. Rather, it is a matter of "bad faith" or as he calls it in French *mauvaise foi* in that it represents when people believe that they are not free due to the fear of the consequences of their choices. Sartre uses the famous example of the waiter to illuminate this alienating condition:

Let us consider this waiter in the cafe. His movement is quick and forward, a little too precise, a little too rapid. He comes toward the patrons with a step a little too quick ... All his behavior seems to us a game ... But what is he playing? We need not watch long before we can explain it: he is playing at being a waiter in a cafe. There is nothing there to surprise us. The game is a kind of marking out and investigation. The child plays with his body in order to explore it, to take inventory of it; the waiter in the cafe plays with his condition in order to realize it. This obligation is not different from that which is imposed on all tradesmen. Their condition is wholly one of ceremony. (Ibid.: 59)

It is thus a "safe" dismissal of our freedom, a willingness to offer our existence up to the whims of fate and circumstance. Returning to a previously discussed concept, it accepts the facticities of one's life as permanent and therefore unalterable.

It is important, though, to better understand exactly what freedom is being surrendered. Doing so reveals a small but profound tension in Sartre's account of existential freedom. If we are doing something freely, then are we genuinely sacrificing our liberty of action? In order to better understand this seeming contradiction, it is necessary to highlight that Sartre is speaking about a specific type or mode of freedom, notably the existential freedom to choose our existence, to explore diverse ways of living and being in the world. It is the difference between "being-in-itself" and "being-for-itself". It is the acceptance of a singular form of Being at the expense of all other forms of existence.

Yet in making this distinction, this raises another perhaps even more difficult problem for existential freedom. It assumes a certain purity in relation to "being-for-itself". There appears an implicit false dichotomy between the "me-as-object" and the completely free from all social influence and determination "being-for-itself". However, if freedom is always embedded within a social context and to an extent culturally defined, then does that mean, in turn, that "being-for-itself" is for intents and purposes impossible? Nevertheless, it also opens up another radical possibility—how to theoretically and practically conceive of an existentially free subject. More precisely, the intentional creation and fostering of social norms, practices, and ethics that promote this deeper form of freedom rather than ones that suppress it.

The Subject of Freedom

While freedom is almost universally presented as a type of liberating force, in practice it most often serves as a means for reproducing a status quo. The popular image of freedom as emancipation from repressive cultural norms and entrenched power relations runs in the face of the concrete ways specific and socially approved types of freedom actually support such domination. Crucial then to the reproduction of power is the ironic but critical reproduction of the free subject. More precisely, it depends on the positive creation and recreation of how individuals and communities experience a sense of agency.

The French thinker Foucault is especially instructive in this regard. For Foucault power is above all else productive in character,

> What makes power hold good, what makes power accepted, is simply the fact that it doesn't only weigh on us as a force that says no, but that it traverses and produces things, it induces pleasures, forms of knowledge, produces discourse. It needs to be considered as a productive network that runs through the whole social body, much more than as a negative instance whose function is repression. (Foucault 1982: 225)

The traditional image of power thus as repressive is undercut by its more dynamic and constructive reality. Rather than attempt to suppress identities and ideas, it instead is focused principally on mass manufacturing them. This productive impulse includes also that of freedom and by association the free subject. Present day freedom is linked to a "self-responsibility and the obligation to maximise one's life as a kind of enterprise" (Rose et al. 2006: 91). Power, in this respect, is at its most powerful not when it is suppressing freedom but conversely when it is actively promoting it.

Consequently, far from being a liberating force, people find themselves having freedom consistently forced upon them. To borrow again from Foucault's terminology, it represents a form of subjection. It reflects the material exercise of power on the body, the socialization of the active subject. He associates this in modern period with the production of "docile bodies" in which

There is no need for arms, physical violence, material constraints. Just a gaze. An inspecting gaze, a gaze which each individual under its weight will end by interiorising to the point that he is his own overseer, each individual thus exercising this surveillance over, and against, himself. A superb formula: power exercised continuously and for what turns out to be at minimal cost. (Foucault 1980: 155)

Here, a specific mode of freedom, like that of market freedom, is a dynamic force that directly acts upon people. To be free, hence, is to live according to proscribed sets of rules and norms that dictate not only actions but also our end goals and the social means through which we pursue them.

It is therefore misguided to associate subjection with a straightforward idea of structuralism or social overdetermination. A key, though underexplored move, within Foucault is that power is fundamentally empowering. Meaning that it is not so much regulatory—asking people for a crude reproduction of its ideals and processes—as it is conducting, opening the space for people to innovate on the basis of its underlying logics and overriding aims (see, for instance, Eskes et al. 1998; Pease 2002). To this end, subjection provides the conditions for people to become subjects of these dominant discourses, crafting and reinforcing a sense of self in relation to (though never completely identical with) these disciplining norms and practices.

Vital to this process of subjectivation is a deep attachment to the perceived freedom ironically linked to subjection. Using other language, the transition from subjection to becoming a subject is one where people maintain a sense of agency in the face of otherwise overdetermining social discourses. According to renowned philosopher Judith Butler, subjection "consists precisely in this fundamental dependence on a discourse we never chose but that, paradoxically, initiates and sustains our agency" (1997: 2). At their most powerful, subjection is strongest when its subjectification revolves around a mode of freedom that reflects its core values and prerogatives. Thus, the free market persists through the production of the "free" subject, whereby paradoxically the regulative demands of the hyper-capitalist society are translated into a conscious embrace of market freedoms as means for realizing one's desires and thus forming one's identity around.

It is here that the affective element of power truly emerges. While Foucault and Lacan are often considered completely opposite in their overall perspectives, recent thinkers have instead stressed their interesting theoretical affinities (see O'Sullivan 2010; Bloom 2015). Specifically, subjectivation involves the psychic investment in a culturally provided identity, akin to Lacan's previously discussed notions of fantasy. Quoting again from Butler, she asks pointedly, "how are we to understand not merely the disciplinary production of the subject but the disciplinary cultivation of an attachment to subjection?" Moreover, she questions

> To what extent has the disciplinary apparatus that attempts to produce and totalize identity become an abiding object of passionate attachment? We cannot simply throw off the identities we become, and Foucault's call to refuse those 'identities' will certainly be met with resistance ... In particular, how are we to understand not merely the disciplinary production of the subject but the disciplinary cultivation of an attachment to subjection? (Butler 1997: 102)

In attempting to answer these questions, Tie (2004: 161) observes:

> [a] subject's complicity in their subjectivation cannot be understood as being purely the effect of their positioning within discourse. Rather, their complicity has an affective dimension. Where a regime of power is able to incite that dimension, it has an increased capacity to become totalising in its effects.

What is often missed is the affective appeal of freedom for this self-creation. It is not just a fantasy that is being invested in—it is an empowering fantasy of freedom linked to capitalist desires for power and control. Drawing on the work of Lacan, modern organizational theorist John Roberts (2005: 637) notes thus that

> [T]he processes of comparison, hierarchization, differentiation, homogenization and exclusion that Foucault observes as the objective mechanisms of discipline have as their necessary correlate similar process "within" the ego as I seek to fix identity in the essentially competitive space of the mirror of my own and other's objectifications.

Dominant modes of freedom, in turn, exist as an illusionary but nonetheless appealing discourse upon which the subject is formed and maintained. Regardless, for instance, of how anxiety producing and disappointing the free market is in practice, the dream of market freedom remains strong and central to how people understand and enact their selfhood.

Revealed, in turn, is a gap of freedom that exists and to a degree persists in the figurative space between subjection and subjectification. It is not a predetermined evolution but a dynamic and necessarily uneven process from which the subject emerges and must be continually socially sustained. It is above all an active and contingent development, captured in the parallel term of "subjectivation". It is in this play of meaning that make possible new forms of social agency and that allows—even if only briefly—a more fundamental existential freedom to appear. This dynamic freedom is witnessed in Foucault's discussion of twentieth-century fitness cultures:

> Mastery and awareness of one's body can be acquired only through the effect of an investment of power in the body: gymnastics, exercises, muscle-building, nudism, glorification of the body beautiful. All of this belongs to the pathway leading to the desire of one's own body, by way of the insistent, persistent, meticulous work of power on the bodies of children or soldiers, the healthy bodies. But once power produces this effect, there inevitably emerge the responding claims and affirmations, those of one's own body against power, of health against the economic system, of pleasure against the moral norms of sexuality, marriage, decency. Suddenly, what had made power strong becomes used to attack it. (1980: 56)

At stake is the role of subjectivation for producing the subject of freedom. Notably, it articulates that which previously was only implicit within subjecting discourses. It places a demand for these norms to express how they reflect and allow for a specific type of freedom. All manifestations of knowledge, therefore, are both instructions for individuals to understand and implement dominant discourses as well as enunciations of how one can be a free subject on the basis of these prevailing norms.

Market knowledge, hence, is an indication of what is required of subjects within a hyper-capitalist system as well as an expression of the empowerment associated with market freedom. Subjects take on a dual meaning—representative of both and at the same time the socialized self-created by a dominant discourse and a cultural statement of what it means to be free. Power, in this regard, is fixated on the subject of freedom. Every instance of power, thus, is an opportunity for re-engaging with this subject of freedom.

Power and Freedom

The neologism that power is everywhere, infusing all aspects of our existence, is in itself a powerful insight. Yet this insight also opens itself up to critiques of having serious ethical deficiencies and even nihilism. It also risks making power invisible and insubstantial—by making it everything it is turned ultimately into nothing or at least nothing distinct. However, these criticisms fail to fully capture the freedom central to this perspective. If power socially produces knowledge and subjects, it also produces existential possibilities for their concrete transformation.

Foucault famously conceives the social as constituted by what he terms "power/knowledge". Power creates, guides, and enacts knowledge for its own reproduction and expansion. While Foucault is attuned to broader systematic influences, his attention rests primarily on the micro-instantiation of "power/knowledge". It is represented in and reinforced by "discourses, institutions, architectural arrangements, regulations, laws, administrative measures, scientific statements, philosophic propositions, morality, philanthropy, etc." (Foucault 1980: 194, 196). At stake is how discourses form and inform subjects in their daily relationships and within existing institutions. Importantly, "What had interested Foucault was not the specific bodies of knowledge compiled through disciplinary investigation at various times. Instead, Foucault had written about the epistemic context within which those bodies of knowledge became intelligible and authoritative" (Rouse 2005: 96).

On first glance, this reading of the social seems worlds apart from existentialism. Yet, digging deeper there emerges potential symmetries. Notably

Foucault recognizes that in the existentialist tradition developed in the twentieth century, the subject—who is, at the same time, part of and reflexively detached from the world 'already there'—is for the first time the real focus of the analysis not only in philosophy but also in social theory. The subject is no longer an ontological being related to sensations or to pure cognitive reason; instead, he/she is always 'in a situation', a part of the world and of the conditions of this world, so that the relationship of the subject to the situation is always open and unpredictable. (Rebughini 2014: 3)

Notably, on positing that which currently "is" as a socially constructed and by no means exhaustive of what could be. Hence

Foucault also claims that this radical contingency becomes the only situational moment in which the subject can develop marginal emancipation from the inevitability of the processes of subjectivation. The subject is someone who exists in the material world with his/her phenomenological and embodied side: the subject is not transcendental, universal or metacultural. (Ibid.: 3)

Knowledge, for Sartre, is a living reflection of the existential relation of being and nothingness. In his view,

Knowledge appears then as a mode of being. Knowing is neither a relation established after the event between two beings, nor is it an activity of one of these two beings, nor is it a quality of a property or a virtue. It is the very being of the for-itself in so far as this is presence to-; that is, in so far as the for-itself has to be its being by making itself not to be a certain being to which it is present. This means that the for-itself can be only in the mode of a reflection (reflet) causing itself to be reflected as not being a certain being. (Sartre 1956: 174)

The pervasive and in many ways all-encompassing norms, practices, and cultural expectations shape our existence. It is the world we enter into and through our socially guided actions and choices continuously strengthen and recreate. However, it is also precisely the recognition of who we are not, "our intuition" of that our present relation to objects in the world is incomplete and subject to reconsideration.

If knowledge then in part reflects the "situation" of existence, then power is—at least in part—the capacity to change it. The mere fact that it is at its core productive signals is creative heart. Quoting the social theorist Mark Bevir (1999: 67) at length, while Foucault would certainly eschew any notion of the fully autonomous subject who "would be able, at least in principle, to have experiences, to reason, to adopt beliefs, and to act, outside all social contexts", he nevertheless points to a different type of subject of agency who

> exist[s] only in specific social contexts, but these contexts never determine how they try to construct themselves. Although agents necessarily exist within regimes of power/knowledge, these regimes do not determine the experiences they can have, the ways they can exercise their reason, the beliefs they can adopt, or the actions they can attempt to perform. Agents are creative beings; it is just that their creativity occurs in a given social context that influences it.

Knowledge, therefore, is a force that equips people with ever new capabilities, potential resources for transforming ourselves and society. Foucault, for his part, distinguishes the "economic" from the "political" as

> discipline increases the force of the body (in economic terms of utility) and diminishes these same forces (in political terms of obedience). In short, it dissociates power from the body; on the one hand, it turns it into an 'aptitude', a 'capacity', which it seeks to increase; on the other hand, it reverses the course of the energy, the power that might result from it, and turns it into a relation of strict subjection. If economic exploitation separates the force and the product of labour, let us say that disciplinary coercion establishes in the body the constricting link between an increased aptitude and an increased domination. (Foucault 1977: 138)

However, this also points to a radical impulse inexorably associated with power—though it is often easily hidden in plain view from us as subjects. Specifically, by asking what society is currently being produced, we become free to profoundly question what type of world could be produced instead. More to the point, in our "economic" production as "political" subjects, there are always "existential gaps" to ask why are we being made so capable and to different ends could these capacities be used.

Power, in turn, is caught in the recurring battle between freedom and domination. According to Foucault, domination denotes the attempt to stagnate social relations. He describes it as a situation "When an individual or social group manages to block a field of relations of power, to render them impassive and invariable and to prevent all reversibility of movement" (Foucault 1988: 3). Note that domination does not refer here to any intrinsic relation of inequality—as, for instance, there may be good reason for a teacher to temporarily hold authority over a student they are instructing. Though never stated by Foucault explicitly, freedom, by contrast, is found in the attempts to empower oneself in dynamic and subversive ways.

This ongoing struggle between power as freedom and power as domination reflects the fundamental existential tension of Being and existence. It is the difference found in "being-for-itself" and "being-in-itself", respectively. While becoming a subject is unavoidable, as one is always necessarily subjected to a social situation, what is not inevitable is the social ordering and sense of self arising from these prevailing discourses. For this reason, Foucault seeks to evolve the Enlightenment idealization of "self-determination" into a radical ethics of "self-fashioning" (McNay 2013). By exploring new lifestyles, opening oneself to fresh experiences, one becomes open to doing otherwise than the expected and demanded. Thus "In effect, Foucault was engaged in a sort of a political therapy, therapy for himself and for those of his readers interested in expanding the possibilities of personal experience and practice within established social orders" (Sawicki 2005: 380).

In doing so, power/knowledge is redefined as power/freedom. The production of the subject is not in the service of any singular way of Being. Rather, it is directed at expanding what is possible both individually and collectively. Foucault hints precisely at such a freedom in his reconsideration of "the problem" of homosexuals:

> Another thing to distrust is the tendency to relate the question of homosexuality to the problem of "Who am I?" and "What is the secret of my desire?" Perhaps it would be better to ask oneself, "What relations, through homosexuality, can be established, invented, multiplied, and modulated?" The problem is not to discover in oneself the truth of one's sex, but, rather, to use one's sexuality henceforth to arrive at a multiplicity of relationships. (1997)

Thus contained within all instances of power are the seeds of both a dominating Being and the growth of a freer existence.

Freeing the Market Subject

The idea of the free market as a liberating force is being increasingly challenged. However, the traditional critique that it is merely an example of capitalist exploitation par excellent also misses how deeply it is transforming people and society. A key feature of neoliberalism, in this respect, is that it is spreading to all areas of social life, a pervasive discourse for reshaping the entirety of our existence. More precisely, it serves as a powerful social knowledge for configuring our worldviews. Tellingly Fourcade and Healy (2007: 10) refer to the current ways we are disciplined to "see like the market"

> Digital traces of individual behaviors (where classifying instruments define what 'behavior' is, and how it should be measured) are increasingly aggregated, stored and analyzed. As new techniques allow for the matching and merging of data from different sources, the results crystallize—for the individuals classified—into what looks like a supercharged form of capital … Markets have learned to 'see' in a new way, and are teaching us to see ourselves in that way, too.

To this effect, the market is less a set of procedures or values and more a dynamic framework for making sense of reality and governing actions.

Significantly, such invasive and increasingly universal knowledge acts to constantly reproduce us as subjects. To borrow a term originally popularized by the French Marxist philosopher Louise Althusser, through its concrete practices it ideologically interpellates individuals as subjects of capitalism. In his view "[An individual's] ideas are his material actions inserted into material practices governed by material rituals which are themselves defined by the material ideological apparatus from which derive the ideas of that subject" (Althusser 1971: 169). In this way, everyday people are incentivized to be better capitalist, reminded of their responsibilities to be more profitable and efficient, to use their time productively, to channel the market towards ethical ends.

However, such diverse interpellation is by no means straightforward. Instead, it often moves in unpredictable ways with surprising results, such as the use capitalist based technological innovations to produce the possibility of a non-market "sharing economy".

Political scientist James Martel (2017) has recently referred to this process as "misinterpellation". Using a range of instances from the past and present, notably how slaves in Haiti were inspired by the Declaration of the Rights of Man that was not intended for them to revolt against their French masters. However, existentialism points to the potential of a slightly different type of "misinterpellation", one not described by Martel. Instead, it depicts when a discourse catalyses the creation of new types of experiences at odds with its original ideological intention. The values associated with the free market, in turn, foster subversive types of social relations and organization such as those associated with hi-tech cooperatives and "makers-spaces".

Fundamentally, this points to the creation of a quite revolutionary political economy—one premised on existential freedom as opposed to domination. In particular, this is based on an inversion of Foucault's initial depiction of economy and politics discussed in the previous section. These moments of doing otherwise than expected introduced renewed opportunities for building new capabilities for the purposes of experimenting with new ways of being. Specifically, it demands using prevailing discourses to empowering novel abilities and interacting that produce fresh forms of "seeing" the world and interacting.

This radical political economy, moreover, changes the aim and character of disciplining. Conventional disciplining, as previously outlined, represents power not as a coercive sovereignty but as an underlying set of norms, social cues, and established practices that almost unconsciously direct behaviour within various cultural contexts and institutions. Rather, it redirects market ideas and capabilities for completely transformational purposes. Thus the original metaphor of the monster, for instance,—one of terror and demonization—can be radically discursively reconfigured to represent fresh possibilities for being. As such,

A monstrous organization theory might bend history and make it groan by interrogating the monstrous evil upon which contemporary organizations rest. But it might also bend trajectories of history by pursuing a monstrous politics and monstrous ethics that monstruct organization life and open it up to other ways of living, working and organizing. (Thanem 2011: 127)

Doing so involves reclaiming existent knowledge for new ideological and material ends. Hence, the idea of automation conventionally associated with a capitalist logic of efficiency and profit becomes a radical opening for the conceiving of "luxury automated Communism" representing a prospective "post-work society, where machines do the heavy lifting and employment as we know it is a thing of the past" (Merchant 2015: N.P.).

Consequently, it opens the possibilities for intervening against the spread of an oppressive knowledge society. In his later work, Foucault develops power/knowledge into a broader theory of "governmentality". He depicts it as an

> ensemble formed by the institutions, procedures, analyses and reflections, the calculations and tactics that allow the exercise of this very specific albeit complex form of power, which has as its target population, as its principal form of knowledge political economy, and as its essential technical means apparatuses of security. (Foucault 1991: 102–103; also see Gane 2008; O'Malley 1996)

Instead of indicating, therefore, a substantive model for governing subjects, it reflects a more foundational production of individuals as governable. It is the shaping of the "conduct of conduct" a particularized relationship with social norms that is based on the need to conform to overriding and innovatively fulfil overriding social expectations. They are "governed" subjects, in so much, that they are governed by discourses. The free market, hence, despite its ostensible rejection of sovereign power and government intervention, depends on this governable subject. This reliance goes beyond the mere demand that people accept the authority of managers and legitimacy of bureaucratic hierarchies. Latham (2000: 2) refers to the emergence of a "social sovereignty" as

> a way to understand how in later modernity both the state and diverse range of non-state actors of interest to Foucault (such as professionals and experts) can both be central to the governance of an increasingly wide range of social domains.

It also involves the prospect that individuals engage with these ideals and such institutional life as forces of governance rather than freedom.

At stake, then, is the reappropriation of the free market towards the creation of difference as opposed to the simple reproduction of the same. It is the concrete use of norms and practices for the sake of reordering—experimenting with what is to test what may be. The innovative impulse that presently runs through neoliberal disciplining is here developed into the ethos of personal and collective discovery, being empowered not to accept the status quo but creatively transform it. It is to use the productive capacity of power to free the market subject as opposed to reinforcing the free market.

The Subject of Existential Freedom

It is understandably a daunting and worrying realization that power is ubiquitous, weaving throughout very fabric of our lives. Yet this insight also gives birth to an equally exciting counterpoint—if the realities of power are everywhere so are the possibilities of freedom. Every moment, every interaction, every instance in which we are subjected is also an opportunity to test and expand existing social limits, to subvert knowledge and convert it into freedom. Yet doing so also exposes the inequality of this existential impulse. It is more available to some than others, in certain places and times. These existential divisions sadly predictably follow along established racial, ethnic, and sexual lines. Freedom, in this respect, is both a birthright and a privilege. Nevertheless, it persists, universal in its global reach and revolutionary in its scope for fostering social change.

It is in Foucault that one finds perhaps the most optimistic modern thinking—in that every discursive formation, every culturally derived norm and action, gives rise to new possibilities, new implications, new gaps of freedom. More precisely, all manifestations reveal what they are not, and in doing so open up the potential to be free to be otherwise. If Foucault is not normative or critical in the traditional philosophical or theoretical sense, it is only because he is excited to patiently keep his options up to see what fresh freedoms emerge, often in the most unlikely of places and most unpredictable of ways. Hence, if knowledge is presumptuous, Foucault shows us that ethics is radically and eternally alert.

This radical patience points to the possibilities of creating a broader ethics of freedom. Foucault speaks, in this respect, of the ethos of "self-care". By this, he does not mean an embrace of market individualism or a culture of total self-absorption. Instead, it is a gesturing towards a radical relation to the self, one that is neither objectified nor purely subjective. He argues that "It is a matter of acts and pleasures, not of desire. It is a matter of the formation of the self through techniques of living, not of repression through prohibition and law" (Foucault 1997: 89). It is assuming the self as a point of experimentation with different ways of being, a concrete site for escaping existing knowledge and seeking out fresh norms and practices to live by. He relates this radical relation to the self to the Greek's notion of a "living" freedom:

> For the Greeks, [ethos] was the concrete form of freedom; this was the way they problematized their freedom. A man possessed of a splendid ethos, who could be admired and put forth as an example, was someone who practiced freedom in a certain way … Extensive work by the self on the self is required for this practice of freedom to take shape in an ethos that is good, beautiful, honorable estimable, memorable and exemplary. (Foucault 1997: 29)

It also opens a novel and arguably more profound point for critiquing persistent inequalities. It places front and centre the question of who gets to experiment with their self, to explore and discover fresh forms of being, and therefore who does not. In this regard, it asks who, in fact, is the subject of freedom and why is this historically the case? In Foucault's own words,

> But above all, one sees that the focus of critique is essentially the cluster of relations that bind the one to the other, or the one to the two others, power, truth and the subject. And if governmentalization is really this movement concerned with subjugating individuals in the very reality of a social practice by mechanisms of power that appeal to a truth, I will say that critique is the movement through which the subject gives itself the right to question truth concerning its power effects and to question power about its discourses of truth. Critique will be the art of voluntary inservitude, of reflective indocility. (Foucault 1997: 386)

It also makes such inequality a subject of freedom rather than merely equality. It intellectually interrogates and practically motivates how those that are excluded from this deeper freedom can regain it on their own terms and as relative to their own specific needs and desires. In decolonizing the knowledge of freedom, one is also decolonizing who it privileges.

Critically, this ethical perspective challenges the power/knowledge framework of liberalism and neoliberalism for a prioritizing instead of power/freedom. On a personal level, it does not demand that people develop their capabilities according to the wants and expectations of prevailing discourses. Nor does it produce them in such a way that they only experience a dominant form of freedom. Conversely, it provides individuals with a social canvas in which to continually repaint their lives, a full-bodied laboratory to experiment with diverse types of possible existences. The innovative social scientist Bent Flyvbjerg calls for a "progressive phronesis" that uses these critical methods to practically intervene and transform our environment. This is not to imply though that such a Foucaultian inspired existential ethics is ever so intentional, it is rather responsive and attentive. It involves a fascination with the opportunities for doing otherwise that power constantly allows, how that which is culturally assumed can be "misinterpellated" in constantly surprising and liberating directions. In this respect,

> the most general set of poles to be found in Foucault's work is the tension between constraining limitations and limitless freedom. Between them there is work on enabling limits. On the one hand, resentment of our limitations can be overcome by recognizing that we are indebted to our constraints. Lives, works of arts and political communities have shape because of constraints. Limitations are ... conditions of possibility. However, to accept given limitations as that which determines all that is possible would make being unbearably heavy. Limits are truly enabling when having given something its form (such as the self), the form engages with its own limits to fashion its own style. (Simons 1995: 3)

Beyond the individual, this ethics also has potentially profound social implications. It jettisons a longing for either an essentialist ethnic or national identity or mere tolerance of an essentialized other. Instead, it champions existential learning, what we individually and as dynamic communities can gain from the ways of being exhibited and discovered by others. It exchanges power and knowledge, in this sense, for shared existences and freedoms. It additionally implies a radical and at times revolutionary existential solidarity—an ethical and political commitment to support all those denied such deeper agency with the freedom to transcend the tyranny of existent knowledge for new ways of being in the world.

The free market, therefore, is a jumping-off point for the detection and engagement with different types of freedom. It also invokes, in turn, a novel form of power. Foucault links current neoliberalism explicitly with what he terms biopower referring to

> The adjustment of the accumulation of men to that of capital, the joining of the growth of human groups to the expansion of productive forces and the differential allocation of profit, were made possible in part by the exercise of biopower in its many forms and modes of application. (1990: 140–41)

This emphasis on health and culturally maintaining the good life reflects the free market's dual desires for sustaining a population that is healthy enough to exploit and wholly supportive of the expansion of its hyper-capitalist culture. Instead, it points to the fostering of a "free life" where one is emancipated from the limits of existing Being and the social disciplining of the pre-given situation of their life. Consequently,

> And that is why, in the end, there can be no such thing as a sad revolutionary. To seek to change the world is to offer a new form of life-celebration. It is to articulate a fresh way of being, which is at once a way of seeing, thinking, acting, and being acted upon. It is to fold Being once again upon itself, this time at a new point, to see what that might yield. (May 2005: 529)

Put forward is nothing less than the existential creation of the subject of existential freedom. We should obviously be under no illusion that such a radical ethics is easily realized—the barriers of knowledge and the power of domination are not so quickly or effortlessly unsettled nor dislodged. However, it does stand as an abiding opportunity and guiding principle for placing existence before Being. It is our best hope of breaking free from our capitalist subjection, exchanging the free market subject for the possibilities of becoming an existentially free subject.

References

Althusser, L. (1971). *For Marx*. London: Verso.

Bevir, M. (1999). Foucault and Critique: Deploying Agency Against Autonomy. *Political Theory, 27*(1), 65–84. https://doi.org/10.1177/009059 1799027001004.

Biebricher, T. (2015). Neoliberalism and Democracy. *Constellations, 22*(2), 255–266. https://doi.org/10.1111/1467-8675.12157.

Blomgren, A.-M. (1997). Nyliberal politisk filosofi. En kritisk analys av Milton Friedman, Robert Nozick och F. A. Hayek. Nora: Bokförlaget Nya Doxa. Quoted in Thorsen, D. E. 2010. The neoliberal challenge. What is neoliberalism? *Contemporary Readings in Law and Social Justice, 2*(2), 188–214.

Bloom, P. (2015). Cutting Off the King's Head: The Self-Disciplining Fantasy of Neoliberal Sovereignty. *New Formations, 88,* 11–29.

Bloom, P. (2017). *The Ethics of Neoliberalism: The Business of Making Capitalism Moral*. Taylor & Francis.

Bloom, P., & Rhodes, C. (2016). Political Leadership in the 21st Century: NeoLiberalism and the Rise of the CEO Politician. In J. Storey, J. Hartley, P. Denis, P. Hart, & D. Ulrich (Eds.), *The Routledge Companion to Leadership* (pp. 359–372). London: Routledge.

Bloom, P., & Rhodes, C. (2018). *The CEO Society: The Corporate Takeover of Everyday Life*. London: Zed Books.

Butler, J. (1997). *The Psychic Life of Power*. Stanford, CA: Stanford University Press.

Davies, W. (2014). *The Limits of Neoliberalism*. London: Sage.

Davies, W. (2015). *The Happiness Industry*. London: Verso.

Eskes, T., Duncan, M., & Miller, E. (1998). The Discourse of Empowerment: Foucault, Marcuse, and Women's Fitness Texts. *Journal of Sport and Social Issues, 22*(3), 317–344. https://doi.org/10.1177/019372398022003006.

Fiet, J. (2001). The Theoretical Side of Teaching Entrepreneurship. *Journal of Business Venturing, 16*(1), 1–24. https://doi.org/10.1016/s0883-9026(99)00041-5.

Foucault, M. (1977). *Discipline and Punishment*. New York: Penguin Books.

Foucault, M. (1978). *The History of Sexuality*. New York: Vintage Books.

Foucault, M. (1980). *Power/Knowledge*. New York: Pantheon.

Foucault, M. (1982). The Subject and Power. *Critical Inquiry, 8*(4), 777–795. https://doi.org/10.1086/448181.

Foucault, M. (1988). The Ethic of Care for the Self as a Practice of Freedom: An Interview with Michel Foucault on January 20, 1984 in the fi nal Foucault: Studies on Michel Foucault's Last Works. *Philosophy & Social Criticism, 12*(2–3), 112–131.

Foucault, M. (1990). *Politics, Philosophy, Culture: Interviews and Other Writings, 1977–1984*. London: Routledge.

Foucault, M. (1991). Governmentality. In G. Burchell, C. Gordon, & P. Miller (Eds.), *The Foucault Effect*. Chicago: University of Chicago Press.

Foucault, M. (1997). What Is the Enlightenment? In P. Rabinow (Ed.), *Ethics: Subjectivity and Truth* (Vol. 1). London: Penguin Books.

Fourcade, M., & Healy, K. (2007). Moral Views of Market Society. *Annual Review of Sociology, 33*(1), 285–311.

Friedman, M. (2009). *Capitalism and Freedom*. Charlottesville: University Press of Virginia.

Gane, M. (2008). Foucault on Governmentality and Liberalism. *Theory, Culture & Society, 25*(7–8), 353–363.

Kreft, S. F., & Sobel, R. S. (2005). Public Policy, Entrepreneurship, and Economic Freedom. *Cata Journal, 25*, 595.

Latham, R. (2000). Social Sovereignty. *Theory, Culture & Society, 17*(4), 1–18. https://doi.org/10.1177/02632760022051284.

Levi-Faur, D., & Jordana, J. (2005). The Rise of Regulatory Capitalism: The Global Diffusion of a New Order. *The Annals of the American Academy of Political and Social Science, 598*(1), 200–217. https://doi.org/10.1177/0002716204273623.

Lukacs, G. (1971). *History and Class Consciousness: Studies in Marxist Dialectics*. London: Merlin Press.

Martel, J. (2017). *The Misinterpellated Subject*. Durham: Duke University Press.

May, T. (2005). To Change the World, to Celebrate Life: Merleau-Ponty and Foucault on the Body. *Philosophy & Social Criticism, 31*(5–6), 517–531. https://doi.org/10.1177/0191453705055487.

McNay, L. (2013). *Foucault and Feminism*. Cambridge: Polity Press.

Merchant, B. (2015). Fully Automated luxury Communism. *The Guardian*.

O'Malley, P. (1996). Risk and Responsibility. In A. Barry, T. Osbourne, & N. Rose (Eds.), *Foucault and Political Reason*. Chicago, IL: University of Chicago Press.

O'Neil, C. (2017). *Weapons of Math Destruction*. New York: B/D/W/Y Broadway Books.

O'Sullivan, S. (2010). Lacan's Ethics and Foucault's Care of the Self': Two Diagrams of the Production of Subjectivity (and of the Subject's Relation to Truth). *Parrhesia, 10*, 51–73.

Ortner, S. (1995). Resistance and the Problem of Ethnographic Refusal. *Comparative Studies in Society and History, 37*(1), 173. https://doi.org/10.1017/s0010417500019587.

Pease, B. (2002). Rethinking Empowerment: A Postmodern Reappraisal for Emancipatory Practice. *British Journal of Social Work, 32*(2), 135–147. https://doi.org/10.1093/bjsw/32.2.135.

Pofeldt, E. (2014). Entrepreneurship Is the Ticket to Freedom. *Forbes*.

Rebughini, P. (2014). Subject, Subjectivity, Subjectivation. *Sociopedia*.

Reich, R. (2009). How Capitalism Is Killing Democracy. *Foreign Policy*.

Rhodes, C., & Bloom, P. (2012). The Cultural Fantasy of Hierarchy: Sovereignty and the Desire for Spiritual Purity. In *Reinventing Hierarchy and Bureaucracy–from the Bureau to Network Organizations* (pp. 141–169). Emerald Group Publishing Limited.

Roberts, J. (2005). The Power of the 'Imaginary' in Disciplinary Processes. *Organization, 12*(5), 619–642.

Rose, J., O'Malley, P., & Valverde, M. (2006). Governmentality. *Annual Review of Law and Social Science, 2*(1), 83–104.

Rouse, J. (2005). Power/Knowledge. In G. Gutting (Ed.), *The Cambridge Companion to Foucault* (pp. 95–112). Cambridge: Cambridge University Press.

Sartre, J. (1956). *Being and Nothingness*. New York: Gallimard.

Sawicki, J. (2005). *Disciplining Foucault: Feminism, Power, and the Body*. London: Routledge.

Simons, J. (1995). *Foucault & the Political*. London: Routledge.

Thanem, T. (2011). *The Monstrous Organization*. Cheltenham: Edward Elgar.

Thorsen, D. (2010). The Neoliberal Challenge: What Is Neoliberalism? *Contemporary Readings in Law and Social Justice, 2*(2), 188–214.

Tie, W. (2004). The Psychic Life of Governmentality. *Culture, Theory and Critique, 45*(2), 161–176. https://doi.org/10.1080/1473578042000283844.

7

Deconstructing the Free Market: The Spectre of Existential Freedom

If the late twentieth century was the age when the market was meant to free us, the twenty-first century has become the era where we are imprisoned by it. It appears as an inescapable social reality from which there is no escape. It defines our present generation, setting limits on our social possibilities, dashing its hopes for collective progress. All that is left is a weary resignation that human potential has truly reached the end of its history.

And yet the new millennium has become, if nothing else, a time of profound change. The anti-establishment politics that are quickly defining the present era represent a mass desire and demand for something radically different. In a recent *New Statesman* article describing the growth of "anti-establishment" politics throughout Europe, the author Mark Leonard (2014: N.P.)—the co-founder and director of the European Council on Foreign Relation—declares

> The bigger problem is an overwhelming sense that our globalised elites have broken free from national loyalties, leaving the middle classes struggling to make ends meet in nation states they no longer control. Until mainstream politicians learn how to grapple with that feeling, they will not be able to tackle the anti-politics mood that is sweeping Europe.

© The Author(s) 2018
P. Bloom, *The Bad Faith in the Free Market*,
https://doi.org/10.1007/978-3-319-76502-0_7

145

Across the ideological spectrum and globe, people are expressing their longings for genuine transformation and their fears over the consequences of these dramatic changes. The failures of the free market and its narrow market freedoms have placed humanity once again at an existential crossroads.

Nevertheless, this deeper freedom has its roots in the very capitalist system which is under existential threat. Historically, capitalism has always careened between presenting itself as ideological unassailable and when necessary eminently reformable. Indeed, the current Conservative Prime Minister of the UK Theresa May has promised to "reform capitalism so that it works for everyone, not just the privileged few". Even top financial executives admit the need for serious changes to the system. Writing in the *Harvard Business Review*, Dominic Barton (2011: N.P.), the global managing director of McKinsey & Company, declares

> First, business and finance must jettison their short-term orientation and revamp incentives and structures in order to focus their organizations on the long term. Second, executives must infuse their organizations with the perspective that serving the interests of all major stakeholders—employees, suppliers, customers, creditors, communities, the environment—is not at odds with the goal of maximizing corporate value; on the contrary, it's essential to achieving that goal. Third, public companies must cure the ills stemming from dispersed and disengaged ownership by bolstering boards' ability to govern like owners.

Critically Marx refers to the "reactive" character of capitalists who are willing to make small changes in reaction to the revolutionary demands of the working class (see Choat 2016). However, contained within this recurring emphasis on reform is the agency of individuals and society to shape the market according to their own economic prerogatives and moral wishes. Thus a relatively constant feature of capitalism is an existential choice as to how its principles should be properly realized in practice.

Indeed, despite its consistent core principles of wage labour and the demand for profit, capitalism is historically marked by its cultural diversity. According to Sanyal (2014: 8), "capitalism's strength and power lies in its ability, not to annihilate its 'other' but to negotiate its world of difference". Similarly, Pasquino (1991: 107) notes that the market exists within "polymorphous universe" where capitalist accumulation is simply one of several discursive powers. This relative diversity is reflected in the "varieties of capitalism" ranging from very pro-market liberal regimes to public-oriented social democracies (see Hall and Soskice 2004). To a certain extent, these represent different variants of capitalist existence.

The rise of neoliberalism has sought to eliminate this admittedly limited existential component of capitalism. Conversely, it has posited capitalism as a singular way of life. Quoting social critic George Monbiot (2016: N.P.)

> So pervasive has neoliberalism become that we seldom even recognise it as an ideology. We appear to accept the proposition that this utopian, millenarian faith describes a neutral force; a kind of biological law, like Darwin's theory of evolution. But the philosophy arose as a conscious attempt to reshape human life and shift the locus of power.

It demands that individuals, communities, countries, the entire world conform to its exclusive vision of existence. It is an essentialized version of capitalist reality, the only form of Being available.

To this end, its survival rests on a logic of perfectibility. Any and all social problems are attributed to an incorrect application of correct free market principles. Hence, what is absolutely key is "getting institutions right" (Choudhary 2007). Former United Nations Secretary-General Kofi Annan proclaimed thus, "good governance is perhaps the single most important factor in eradicating poverty and promoting development". If there is a fault, it is with people, not the system or its ideals. Rather, the blame lies with individuals and governments who refuse to be rational and accept the supposed self-evident truth of neoliberalism.

These demands for the perfect market subject, though, have been undercut by the sheer visible and statistical evidence of capitalism's current failings. Now it is the free market itself that is perceived the need to be perfected. It must be consistently and innovatively applied to address social and economic ills, many of which are directly attributed to the spread of the free market itself. The legitimacy of the free market, therefore, increasingly rests on its flexibility for transforming a society it created.

There is obviously a profound danger in relying on or even expecting the free market to ultimately fix itself. Yet there is a deeper and more radical trend emerging, in which the free market has at best a very limited place. There is a creative spirit appearing, slowly perhaps, but surely, aimed at experimenting with new forms of social organization and relations, infused with a desire to expand what is possible in the hopes of better embodying shared desires for equality, democracy, and freedom.

Here the capitalist imperative for innovation is exchanged for a more radical spirit of deconstruction. It is one premised on subverting existing orderings so that they can be reconstructed in new and exciting ways. Perfectibility is transformed into a force for pushing this deconstruction ever further, for fermenting in people an insatiable desire to play with prevailing cultural grammars to explore fresh social configurations that come closer to assuring us of material security and shared freedom. This eternal promise of a freedom to come haunts the free market, existing as a spectre foretelling its ultimate deconstruction.

Rewriting Freedom

The free market portends limitless possibility for individuals. Central to its appeal is a mythology of infinite upward mobility for all those with the determination and talent to succeed on their own merits. Moreover, it contrasts itself with other social systems by promising people the ability to pursue the dream of their choosing. Nevertheless, these personal aspirations are quite socially constructed, their goals culturally scripted to reflect capitalist values and the triumph of market freedom.

Significantly, the story of freedom is always being written. While it is often theorized as intrinsic, freedom nevertheless has a long documented history. It is not just a matter of being recorded for posterity either. Indeed, modern conceptions of liberty can be directly traced back to the founding texts of Locke, Rousseau, and Montaigne among others. Further, the text of liberty, it being put down in words and publicly disseminated, serves as a force for teaching people how to be free. Thus in the US, perhaps the country where the value of freedom is most "cherished" in principle, if not always in practice, its

> accounts of freedom tend to be historical rather than theoretical. Freedom has provided the most popular "master narrative" for accounts of our past, from textbooks with titles like *Land of the Free* to multivolume accounts of the unfolding of the idea of freedom on the North American Continent. Such works, while valuable in situating the experience of freedom in historical context, tend to give it a fixed definition and then trace out how this has been worked out over time. Generally, they ground American freedom in ideas that have not changed essentially since the ancient world, or in forms of constitutional government and civil and political liberty inherited by England and institutionalized by the founding fathers. (Foner 1999: xiv)

Indeed, it is worth noting that freedoms are commonly enshrined through constitutions. They are quite literally socially constituted in this respect. Whereas the free market does not have a constitution as such, it certainly has modern-day gospels such as Hayek's *The Road to Serfdom* and Milton Friedman's *Capitalism and Freedom*. This text supposedly uncovers the "truth" of freedom, spelling out its inexorable relationship to a market existence. In practice, these dogmatic tenets of market freedom have been reinforced through mythical narratives. These dream scenarios are only strengthened by martial depictions of these freedoms being under threat and needing defending. Tellingly, such narratives shape how people conceive and experience such freedom. They act as founts of cultural knowledge, fictionalized accounts giving rise to concrete realities. In the words of Gordon Gekko in the famous 1980s movie *Wall Street*

> I am not a destroyer of companies. I am a liberator of them! The point is, ladies and gentleman, that greed—for lack of a better word—is good. Greed is right. Greed works. Greed clarifies, cuts through, and captures the essence of the evolutionary spirit. Greed, in all of its forms—greed for life, for money, for love, knowledge—has marked the upward surge of mankind.

Market freedom thus follows a regular pattern, an accepted narrative of personal agency and success. There is an approved story of freedom, one that reflects the idealized capitalist values of hard work, determination, and talent. It is the nineteenth-century American story of Horatio Alger who "pulled himself up by his bootstraps" to success. Under neoliberalism this script has been infused by a spirit of entrepreneurism. Indeed,

> The older and still dominant American myth involves two kinds of actors: entrepreneurial heroes and industrial drones—the inspired and the perspired. In this myth, entrepreneurial heroes personify freedom and creativity. They come up with the Big Ideas and build the organizations—the Big Machines—that turn them into reality. They take the initiative, come up with technological and organizational innovations, devise new solutions to old problems. They are the men and women who start vibrant new companies, turn around failing companies, and shake up staid ones. To all endeavours they apply daring and imagination. (Reich 1987)

These scripts have been further reinforced by public to employability and "self-improvement" as the ultimate pathways to individual freedom.

The present-day scripting of freedom is inexorably linked to text-based forms of contemporary capitalist disciplining. The increased presence of mobile technology has digitalized and customized this overarching story of market freedom. There are now apps for every desire, for all addressing perceived weaknesses. They create personal narratives uniquely crafted to maximize our individual value—making us more efficient and healthy subjects. To this extent, technology represents

> …the 'good' power to increase dramatically our productivity and social capital, become our life recorder, or help us organize a rally. The flip side of this is the belief that mobile technologies are powerfully bad, inciting us to riot, affray, excessive sociability or solipsism, or crimes against grammar or cultural values. (Goggin 2006: 206–207)

This personalized textual disciplining is, furthermore, upscaled to the collective level. Economic downturns and social ills are due to the personal weakness of its citizens and as such the state. The financial crisis was thus the result of a greedy public and their profligate government, demanding therefore strict austerity policies. National statistics and budgets are constant reminders if populations are keeping on the right track.

It is tempting to try to resist these quite literal inscriptions of capitalist power through simply writing a new story of freedom. To counter the hegemony of market freedom, fresh stories must be told and novel constitutions enshrined. While undeniably important, more radical measures are also called for. Namely, we must change how we approach and engage with the very "text" of our lives and in doing so begin completely rewriting the possibilities of our freedom and existence.

Creating a Collective Life Project

A central question of contemporary existence is how to lead a meaningful life. While such concerns are perhaps timeless in their relevance, it has arguably especially significant in this day and age. The rise of neoliberalism brings with it further complications as it reduces the entirety of human existence to economic values. However, the market freedoms it instils are meant to provide people with the knowledge and skills necessary to achieve personal fulfilment as well as professional success. It reframes the question into an optimistic and instrumental story of constant upward mobility with the aim of living well.

Sartre for his part introduces his own radicalized version of existential self-actualization. He proposes the creation of what can be termed "life projects" for cultivating a freer mode of existence. He argues,

> Thus in what we shall call the world of the immediate, which delivers itself to our unreflective consciousness, we do not first appear to ourselves, to be thrown subsequently into enterprises. Our being is immediately "in situation"; that is, it arises in enterprises and knows itself first in so far as it is reflected in those enterprises. We discover ourselves then in a world peopled with demands, in the heart of projects "in the course of realization". (Sartre 1956: 39)

To this effect,

My consciousness is not restricted to envisioning a negative. It constitutes itself in its own flesh as the nihilation of a possibility which another human reality projects as its possibility. For that reason it must arise in the world as a Not; it is as a Not that the slave first apprehends the master, or that the prisoner who is trying to escape sees the guard who is watching him. (Sartre 1956: 47)

Notably, it is about coordinating the "in-itself" and "for-itself" aspects of being—utilizing one's freedom to transform their facts in accordance with one's desires and actions. Returning to Sartre (1956: 102),

What must be noted here is that the For-itself is separated from the Presence-to-itself which it lacks and which is its own possibility, in one sense separated by Nothing and in another sense by the totality of the existent in the world, inasmuch as the For-itself, lacking or possible, is For-itself as a presence to a certain state of the world. In this sense the being beyond which the For-itself projects the coincidence with itself is the world or distance of infinite being beyond which man must be reunited with his possible.

In this respect, self-actualization is a process of choosing quite literally one's own life as much as is possible.

However, this life project is far from such a straightforward venture. Indeed, while the free market promises constant improvement, as hard work and perseverance lead one inevitably towards one's goals, Sartre describes a much more ambiguous and complicated reality. He recognizes that such self-actualization in the final analysis is necessarily impossible. Hence, "Perception is naturally surpassed toward action; better yet, it can be revealed only in and through projects of action. The world is revealed as an 'always future hollow', for we are always future to ourselves" (Ibid.: 322). There is always a certain incoherence to such attempts at coordinating what is for what one would like to be.

Despite this disjointedness there remains, in Sartre's view, a deep and persistent "desire for being". While he rejects any original or transcendental grounding, he does subscribe to a certain metaphysics of the consciousness that leads naturally to a longing on the part of the self to be. This conscious desire can take on several manifestations—(1) "for-itself" turning into an "in-itself", (2) "for-itself" seeking to be an all-powerful godlike "for-itself", and (3) "for-itself-in-itself". These represent different fundamental projects, one based on an original choice—often not fully known to the individual themselves—that has shaped their relationship to freedom. Sartre, therefore, rails against the inauthentic choice of basing your "life project" solely on what is provided to you by others in the world, what he terms "being-with":

> And it is very true that I am responsible for my being-for the Other in so far as I realize him freely in authenticity or in unauthenticity. It is in complete freedom and by an original choice that, for example, I realize my being-with in the anonymous form of "they." And if I am asked how my "being-with" can exist for-myself, I must reply that through the world I make known to myself what I am. In particular when I am in the unauthentic mode of the "they," the world refers to me a sort of impersonal reflection of my unauthentic possibilities in the form of instruments and complexes of instruments which belong to "everybody" and which belong to me in so far as I am "everybody". (Sartre 1956: 246)

This reading of course raises immediate questions of whether one can ever lead an authentic and meaningful existentialist life. The key, for doing so, is to ironically embrace rather than merely despair at this inherent ambiguity of conscious existence. According to modern existentialist philosopher Linda Bell (1989: 45), a person "would be right if he recognized himself as a being that is what it is not and is not what it is" and therefore what is required is "the awareness and acceptance of—this basic ambiguity" (Bell 1989: 46). In his later theorizing, Sartre compares to having a "lucid recognition" of these tensions and acting accordingly. He writes "Authenticity, it is almost needless to say, consists in having a true and lucid consciousness of the situation, in assuming the responsibilities and risks it involves, in accepting it … sometimes in horror and hate" (1948: 90). To this end, it involves continually and critically "willing one's freedom", transforming their situation into an opportunity for choosing an existence and as such self that one desires.

In this way, with every choice one is figuratively writing their own story of existence. It is the application of an infinite and universal freedom to a contingent and relative concrete reality. He argues that in considering the relation between belief and action:

There is no doubt that I could have done otherwise, but that is not the problem. It ought to be formulated like this: could I have done otherwise without perceptibly modifying the organic totality of the projects that make up who I am? … instead of remaining a purely local and accidental modification of my behavior, it could be effected only by means of a radical transformation of my being-in-the-world … In other words: I could have acted otherwise. Agreed. But at what price? (Sartre 1956: 454)

While these insights have undeniable practical implications, they are part and parcel of a broader dialectic—albeit one that is never fully synthesized. Specifically, Sartre distinguishes "praxis", denoting human activity in the material world, from the "practico-inert" the ways we are influenced by institutional and normative histories. Indeed in his later *Search for a Method*,

[I]t is not true that History appears to us as an entirely alien force. Each day with our own hands we make it something other than what we believe we are making it, and History, backfiring, makes us other than we believe ourselves to be or to become. Yet it is less opaque than it was. (Sartre 1963: 90)

To this end, the application of praxis—such as in speaking—transforms and is limited by the "practico-inert" force of language. On this basis, contemporary existentialist philosopher Paul Gyllenhammer (2015: 8) argues for a "progressive" notion of the "practico-inert" noting that

Underlining Sartre's moral judgments is his steadfast commitment to the freedom and, hence, uniqueness of each individual (this goes back to the correlated issues of nominalism and atheism). Individuals can maintain a level of creative freedom even within the limits of a social (even biological) medium. Thus, Sartre's moral views are governed by (1) the individual's struggle to maintain a distance from her belonging to a socially prescribed normative horizon and (2) the respect individuals owe to the dignity of other individuals. In a Kantian sense, respect is derived from the unique power all humans have to be creative beings. And our practical responsibility is governed by each person's creative power to resist the actual, oppressive world and to work for the possible, integral society. (Given this role of the possible, I maintain that Sartre's view is ultimately based on hope.)

Self-actualization is found in this ongoing interaction between "praxis" and the "practico-inert", our use and reuse of the existing world and its structures.

Whilst certainly compelling for the individual, this perspective appears to leave little room for collective freedom. Sartre addresses this gap to an extent with his proposal of a "third mediating force" that of the "group". He rejects "others", in this respect, as an ultimately alienating and objectifying force, categorizing people in a reified "us" that restricts their freedom of choice. In "Being and Nothingness", he also views other people as "opportunities and chances" for one's own individual creative freedom—an optimism that is counterposed by his rather infamous sentiment that "hell is other people" in so much as their perceived judgements can unduly shape an individual's willingness to be free. Nevertheless, he does introduce the further concept of "we", the recognition of the freedom found in a group that creates their own norms and social orderings.

This conception of a free "we" as opposed to an objectified "us" serves as a foundation for the creation of a more radical collective life project. Critically building on Sartre's insights, Beauvoir (1948) proposes a rather revolutionary idea of existentialist authenticity. She argues that in willing one's own freedom, one is in fact willing such freedom for everyone. Importantly, in her view,

> just as life is identified with the will-to-live, freedom always appears as a movement of liberation. It is only by prolonging itself through the freedom of others that it manages to surpass death itself and to realize itself as an indefinite unity. (Ibid.: 32)

This invokes a solidarity between people based on their shared realization of "engaged freedom", in which personal agency is necessarily linked to the creation of a future open to possibility generally. In her words,

> by turning toward this freedom we are going to discover a principle of action whose range will be universal. The characteristic feature of all ethics is to consider human life as a game that can be won or lost and to teach man the means of winning. Now, we have seen that the original scheme of

man is ambiguous: he wants to be, and to the extent that he coincides with this wish, he fails. All the plans in which this will to be is actualized are condemned; and the ends circumscribed by these plans remain mirages. Human transcendence is vainly engulfed in those miscarried attempts. But man also wills himself to be a disclosure of being, and if he coincides with this wish, he wins, for the fact is that the world becomes present by his presence in it. But the disclosure implies a perpetual tension to keep being at a certain distance, to tear one self from the world, and to assert oneself as a freedom. To wish for the disclosure of the world and to assert oneself as freedom are one and the same movement. Freedom is the source from which all significations and all values spring. It is the original condition of all justification of existence. The man who seeks to justify his life must want freedom itself absolutely and above everything else. At the same time that it requires the realization of concrete ends, of particular projects, it requires itself universally. (Ibid.: 23)

Thus, she proclaims "My freedom must not seek to trap being but to disclose it. The disclosure is the transition from being to existence. The goal which my freedom aims at is conquering existence across the always inadequate density of being" (Ibid.: 30). At stake in the creation of such a collective life project is the freedom to deconstruct and reconstruct our existing social narratives, discovering fresh interpretations and potentials for being in the world.

The Play of Freedom

Existentialism is often accused by its critics of being fundamentally despairing—a depressing account of human existence where we are in Sartre's own words "condemned to freedom". Yet such a negative view is only relevant if one ignores Sartre's broader intellectual project, namely the stripping of all philosophy of its "seriousness". His goal, perhaps above all others, is to create a new framework for thinking and acting that rejects any internal essences or external authority. Instead, he wants to philosophize about how humans can more fully exist within the eternal play of their own freedom. In Sartre's view,

To be sure, it must be noted first that play as contrasted with the spirit of seriousness appears to be, the least possessive attitude; it strips the real of its reality. The serious attitude involves starting from the world and attributing more reality to the world than to oneself; at the very least the serious Man confers reality on himself to the degree to which he belongs to the world. (1956: 580)

To do so invariably involves a reflective perspective on one's one life and choices. As previously highlighted, our decisions are not made in a vacuum nor are they predetermined by the situation in which we are born into. Rather they are cumulative reflections of our previous choices, each one transforming and reinforcing the interpretation of our contingent material and social reality.

If the given cannot explain the intention, it is necessary that the intention by its upsurge realize a rupture with the given, whatever this may be. Such must be the case, for otherwise we should have a present plenitude succeeding in continuity a present plenitude, and we could not prefigure the future. Moreover, this rupture is necessary for the appreciation of the given. The given, in fact, could never be a cause for an action if it were not appreciated [given a meaning by consciousness]. But this appreciation can be realized only by a withdrawal in relation to the given, a putting of the given into parentheses, which exactly supposes a break in continuity. (Sartre 1956: 478)

Sartre, therefore, advocates a form of "self-transformation" revolving around a type of hermeneutic critical reflection in which we trace back our choices and their meanings to discover their origins and initial motivations (see Busch 2002). It is a means of looking backwards in order to more freely move forwards, following our past trajectories so as to more freely guide our future ones. He notes,

What was once both a vague comprehension of our class, of our social conditioning by way of the family group, and a blind going-beyond, an awkward effort to wrench ourselves away from all this, at last ends up inscribed in us in the form of character. At this level are found the learned gestures (bourgeois gestures, socialist gestures) and the contradictory roles which compose us and which tear us apart (e.g., for Flaubert, the role of the dreamy pious child, and that of the future surgeon, the son of an atheistic surgeon). At this level also are the traces left by our first revolts, our desperate attempts to go beyond a stifling reality, and the resulting deviations and distortions. To surpass all that is also to preserve it. We shall think with these gestures which we have learned and which we want to reject. (Sartre 1963: 101)

This form of reflection speaks to the French philosopher Jacque Derrida's later philosophy of deconstruction. Interestingly, there appears to be no intellectual love lost between these philosophical giants of the twentieth century. Nevertheless, their comparative perspectives have more resonance than is perhaps commonly assumed. Specifically, the existential hermeneutics advocated by Sartre is strengthened by adopting a further deconstructive dimension to such critical reflections. Derrida, in this respect, focuses on the intrinsic differences found in all textual interpretations. Specifically, he emphasizes two as particularly formative—the synchronicity of words, as their meaning is dependent on their relationship with other words, and the diachronic gap between contemporary and historical meanings.

Whilst Derrida is speaking explicitly of textual interpretations, these are not confined to the merely literary. He famously—or infamously as one may have it—observed that "there is nothing outside the text". As such they represent our interpretive relation with our material and social reality. This chimes with Sartre's own existentialism based ultimately on our constant conscious interpretations of our situated reality. To this effect, the world is not given but rather humanly interpreted and these interpretations actively shape our understanding of our given situation and the facts of our existence. Freedom is found not so much in action, which is always to a certain extent confined, but a conscious reading of reality to better reflect our infinite freedom and keep open our possibilities.

The potential contribution of deconstruction to existential freedom is, moreover, apparent in Derrida's broader metaphysics of presence. Here he decries the current reification of what is, those elements of the social that are prioritized and made visible. Instead, he seeks out what is absent in such presence, what story is not being told, what way of viewing the world is being blotted out for these dominant interpretations. Hence, he aims to uncover and bring back to light "the margins of philosophy" those readings and interpretations that

> blur the line which separates a text from its controlled margins. They interrogate philosophy beyond its meaning ... not only to reckon with the entire logic of the margin but also to take on entirely another reckoning: which is doubtless to recall that beyond the philosophic text there is not a blank, virgin empty margin, but another text, a weave of differences of forces, without any present centre or reference. (Derrida 1982: xxiii)

To this extent, he critically asks:

> Can this text become the margin of a margin? Where has the body of the text gone when the text is no longer a secondary virginity but an inexhaustible reserve, the stereographic activity of an entirely different ear? (Ibid.: xxiii)

This overriding concern has clear resonances with Sartre's own existentialist project. Namely, it is in the critical interpretation of reality that we are able to reject what is for what is not yet. By continually looking at our present choices and placing them within the broader totality of our interpretive existence, we are able to reconstruct, change, and expand our life potentials.

Significantly, at a deeper philosophical level, there is a profound affinity between Sartre's nothingness and Derridean absence. Non-being for Sartre is absolutely fundamental to freedom it is that which rejects a current situation for the possibility of something more. Similarly, by identifying and decrying what is present, Derrida holds out the potential for discovering a greater abundance of meanings in the form of current absences. Consequently, "in every exposition it would be exposed to disappearing as disappearance. It would risk appearing: disappearing" (Ibid.: 6). Both view Being as contingent and subject to interpretation, thus catalysing the need for the perpetual reinterpretation of our given world.

Emerging from this shared interpretive freedom is the radical potential of the collective play of freedom. It is paradoxically in the absence of freedom that new forms of freedom emerge just as it is non-Being the fresh types of being are made possible. We are only liberated from the continual constructions and reconstructions of our reality via our active and intentional deconstruction of it. This echoes Derrida's notion of the "supplement"—that which is always additional, unseen but present even in its absence, to an existing textual object. Quoting him at length

> For the concept of the supplement—which here determines that of the representative image—harbours within itself two significations whose cohabitation is as strange as it is necessary. The supplement adds itself, it is a surplus, a plenitude enriching another plenitude, the fullest measure of presence. But the supplement supplements. It adds only to replace. It intervenes or insinuates itself in-the-place-of; if it fills, it is as one fills a void. If it represents and makes an image, it is by the anterior default of a presence…. The scandal is that the sign, the image, or the representer, become forces and make "the world move". Blindness to the supplement is the law. We must begin wherever we are and the thought of the trace, which cannot take the scent into account, has already taught of the trace, which cannot not take the scent into account, has already taught us that it was impossible to justify a point of departure absolutely, Wherever we are: in a text where we already believe ourselves to be. (Derrida 1974: 147)

It is this play of meaning, creatively interweaving what is with what is not, that which is present with that which is absent, that we are emancipated from the "facts" of our existence. And in maintaining this difference—this existential gap in the changeable and various ways we can interpret our reality—we preserve a constant space for the enaction of our freedom to imagine and make real new realities.

Perfecting Freedom

Existentialism and Deconstruction are fundamentally embedded in the world. Indeed Sartre makes clear that if there is anything essential about existentialism it is that there is nothing beyond or underlying human existence. Similarly, Derrida distinguishes his own brand of deconstruction from a traditional Kantian critique, as he argues that such critical interpretations will always be shadowed by the "dogmatic baggage" of the language it is seeking to meaningfully. Paradoxically, it is this inescapability, this total and complete immersion in the social, that allows for the radical transcendence of existent realities and cultural orderings. By not seeking to go beyond the world, it focuses sole attention on how it can be continually and creatively transformed.

Indeed, the notion of transcendence forms a key part of Sartre's overall existentialist philosophy. Put simply, it is the ability of individuals to exceed their given situation. As such

> we can define action as our being-in-the-world in so far as we have to be it in the form of being-an-instrument-in-the-midst-of-the-world. But if I am in the midst of the world, this is because I have caused the world to-be-there by transcending being toward myself. And if I am an instrument in the world, this is because I have caused instruments in general to-be-there by the projection of myself toward my possibles. (Sartre 1956: 325)

The very fact of our consciousness already implies our capacity for and tendency towards transcendence. More precisely, that we can think and critically view the world beyond the "brute facts" of its giveness. By the very definition provided to it by Sartre, consciousness is "the transcending For-itself" in which "Consciousness is a being such that in its being, its being is in question in so far as this being implies a being other than itself" (Ibid.: 629). This intrinsic transcendence is further developed and activated through our interpretive abilities—as how we choose to understand and makes sense of the world necessarily shapes that which is considered important and predominant.

Initially, it may seem that deconstruction is much less optimistic. Indeed, its philosophical innovation does not lead necessarily or always explicitly to a clear political or normative agenda. However, in his later works, Derrida explores the potential radical implications of his ideas through the notion of perfectibility. He especially links it to a notion of "democracy to come", proclaiming

> ...the inherited concept of democracy is the only one that welcomes the possibility of being contested, of contesting itself, of criticizing and indefi-

nitely improving itself. If it were still the name of a regime, it would be the name of the only regime that presupposes its own perfectibility and thus its own historicity. (Derrida 2004: 24)

In particular, he rejects democracy as a "regulative idea" and instead highlights it as a force for keeping open up the contingency of social relations.

The idea of a promise is inscribed in the idea of a democracy: equality, freedom, freedom of speech, freedom of the press—all these things are inscribed as promises within democracy. Democracy is a promise. That is why it is a more historical concept of the political—it's the only concept of a regime or a political organization in which history, that is the endless process of improvement and perfectibility, is inscribed in the concept. So it's a historical concept through and through, and that is why I call it "to-come". (Derrida 1997)

While at points Derrida overtly supports liberal democracy, for this reason, his concept of perfectibility far exceeds its conventional limits and ideals. Instead it denotes that the social is always necessarily under construction and always striving for to be more perfectly aligned with its highest ideals. It is the "promise" of a freedom always "to come".

This emphasis on perfectibility reverberates with Derrida's concept of erasure. Notably, he commonly refuses to define deconstruction in positive substantive terms. Instead he tellingly refers to what it is not—existing neither as a method, a critique, nor form of analysis. However, he accepts that these terms will at points have to be returned to and exploited to describe deconstruction despite their imperfections. In doing so, he hopes to show their inadequacy, troubling their meanings and ultimately "erasing" the fundamental philosophical assumptions underlying them. He contrasts this with the ways that any assertion is a form of erasure in that it represents "…a body that in turn produces itself by erasing itself as the barely visible, entirely transparent, representation of both the philosophical and sociopolitical corpus, the contract between these bodies never being brought to the foreground" (Derrida 2002a: 90). A radical erasure by contrasts recognizes precisely what is being erased, what is being made absent, what is actually being unsupposed as simply present and neutral. Likewise, by continually returning to conventional definitions of freedom, their insufficiency is shown and new forms of freedom are critically sought out. Thus similar to Sartre, the transcendence of what is given, its "erasure" for something different and new, is made possible through recurring processes of negation and revealment—more precisely, revealing what has been negated so that what is there can be negated (see—especially in the context of teaching—Bingham 2007).

Consequently, freedom is never fully realized for either Sartre or Derrida. More precisely, its vitality is found in its eternal incompleteness. Deconstruction is forever postponed, according to Derrida, ongoing rather than finished (see Derrida 1978). It is a constant process of deferral, and it is this eternal postponement that the subject is constituted and reconstituted. Hence

> It confirms that the subject, and first of all the conscious and speaking subject, depends upon the system of differences and the movement of différance, that the subject is not present, nor above all present to itself before différance, that the subject is constituted only in being divided from itself, in becoming space, in temporizing, in deferral… (Derrida 1981: 28–30)

Consequently, the point of philosophy is to

> …traverse a phase of overturning. To do justice to this necessity is to recognize that in a classical philosophical opposition we are not dealing with the peaceful coexistence of a vis-a-vis, but rather with a violent hierarchy. One of the two terms governs the other (axiologically, logically, etc.), or has the upper hand. To deconstruct the opposition, first of all, is to overturn the hierarchy at a given moment. To overlook this phase of overturning is to forget the conflictual and subordinating structure of opposition. (Ibid.: 42)

Analogously, freedom for Sartre, is unending, infinite, and continually demanding. There is no end point at which it is ever completely exhausted. In this regard, freedom is forever postponed, still on the horizon, waiting to be experienced.

Freedom, hence, can never be made perfect; it is at best always only perfectible. In deconstructing and reconstructing meaning and practices, fresh "opportunities and chances" to experience freedom emerge. There are ever new absences to explore and established realities to unsettle, erase, and transcend. The ability to shape our existence, to move from what is to what could be, is intrinsically partial. As such, for every claim to freedom, there is an equal and powerful chance for its further perfection.

Deconstructing Market Freedom

The free market leaves, it seems, little room for interpretation. Its supporters present it as based on objective and incontrovertible natural economic laws that directly correspond to human nature. However, it is constantly open to question. Popular movements railing against its excesses and failures have shown the possibility and need for its profound reinterpretation. To deconstruct its ideas of meritocracy, competition, and profit in order to ask what is missing in such accounts and what values and types of social order could emerge with their erasure for different ones.

The recent era after the financial crisis has certainly witnessed a general rethinking of neoliberal ideals—popularly challenging the perceived logic behind trickle-down economics, tax cuts for the rich, and the global race to the bottom. However, it is still questionable the extent this has led to a dramatic reconsideration of market freedom. The ideas of employability and marketability have retained their relevance as the primary means for people to experience any sense of existential agency (see Arthur and Rousseau 2001; Hall 2004). Tellingly, attempts to counter these established freedoms usually involve finding ways to cope with unavoidable pressures of an increasingly demanding market society (see Bloom 2017).

A crucial component in the radical reinterpretation of the free market is to highlight its meaningful and therefore ultimately existential violence. More precisely, to reveal how its objectified understandings and legal standing hide its forceful suppression of different social interpretations and realities from emerging. Derrida rather famously theorized the "mystical" and violent foundations of any and all constituted social law. He argues that

Everything would still be simple if this distinction between a justice and law were a true distinction, an opposition the functioning of which was logically regulated and masterable. But it turns out that the law claims to exercise itself in the name of justice and that justice demands for itself that it be established in the name of a law that must be put to work (constituted and applied) by force "enforced". Deconstruction always finds itself and moves itself between these two poles. (Derrida 2002b: 250–251)

Their birth in literal and figurative violence, the aggressive overturning of the old, is given a mystical justification in discourses of legality and essentialism following their victory. Accordingly

Here a silence is walled up in the violent structure of the founding act. The "mystical" is an abyss in the heart of what is supposedly well founded: vanished cruelties at the moment of constituting a state, forgotten terror when new law comes into force, events which remain historically "ininterpretables ou indéchiffrables/uninterpretable or indecipherable". (Ibid.)

Hence the interruptive moment of any change to the status quo is historically masked by a perceived unalterable and divinely given present. To this end, the supposed objective nature and legal rights of market freedom obscure its contingent and intentionally unsettling beginnings.

Recapturing this disruptive revolutionary spirit involves critically assessing and meaningfully subverting the discursive ways market freedom continues to survive and thrive. It is not merely that the free market constitutes a dominant present-day hegemony, marginalizing all other social interpretations; it is important how it is dynamically presencing itself. A key way in which it is currently doing so is by highlighting its own absence. In other words, in the wake of a precarious economy steeped in ever greater material insecurity and psychological anxiety, there is a general acceptance of the need to help people become more employable and marketable to overcome these obstacles. To counter this strategic absence, instead, is to socially draw attention to what types of new personal and social freedoms the free market occludes.

To this end, the free market is sustained by myths of its own perfectibility. Even at its worst, close to a global financial crisis, it can always present itself as still capable of being improved. Thus only a year after the 2008 financial crisis, then President Obama (2009: N.P.) confidently declared

> The answers to our problems don't lie beyond our reach. They exist in our laboratories and our universities; in our fields and our factories; in the imaginations of our entrepreneurs. What is required now is for this country to pull together, confront boldly the challenges we face, and take responsibility for our future once more.

Of course, this presents a substantial contradiction for market freedom—given that in theory it is already perfect. Consequently, what must be perfected are individuals and societies themselves. As such "contemporary tendencies to economize public domains and methods of government also dialectically produce tendencies to moralize markets in general and business enterprises in particular" (Shamir 2008).

It is people that must be made perfectible, constantly maximizing their market value and their ability to negotiate the pressures achieving professional success and personal fulfilment.

The antidote to this market demand for personal perfectibility is to highlight the imperfection, or more precisely incompletion, of freedom under capitalism. It is to reveal how the exclusive association of freedom with the free market leads to the postponement of the emergence of new and even more exciting forms of agency. Critically, this revealment goes beyond the mere abstract, a discursive expansion of the public imagination. Rather it also depends on highlighting new presences of freedom, revealing how various capabilities and types of social relations can breed novel ways of existence and fresh means for shaping our environment.

A fundamental task then for transcending the free market is to radically deconstruct market freedom. It is to constantly reveal in word and practice its absences. To highlight its everyday meaningful violence it performs by preventing different freedoms from manifesting and becoming widely available. It must be stripped of its "mystical" trappings to reflect its contingent and alterable nature. And crucially, it requires bearing witness to the role the free market plays in postponing a more perfect existential freedom.

The Spectre of Existential Freedom

There is a paradox at the heart of freedom. It is both always present and equally always absent. The possibility to make a "free" choice is constantly available, obviously though the conditions for this free choice is consistently lacking. More importantly, every manifestation of freedom hides the potential for and instantiation of another kind. The possibility of freedom thus is eternally present and forever haunting us.

Derrida speaks of just such a phenomenon—especially in relation to Marxism. He refers to it as a "spectre" that continually interjects on our contemporary ideals and pushes to go further in our pursuit of human emancipation (Derrida 1994). This use of the word spectre draws upon Marx and Engel's own famous use of the term in reference to Communism in the opening lines of the Communist Manifesto. A spectre, in this respect, both shadows the current order and foreshadows the new one to come. The promise of existential freedom similarly challenges the liberty of a current social order while pointing the way to a more liberated future.

The danger, of course, is that this future will be influenced and itself haunted by the history and ideological content of such a spectre before it has even begun. To avoid, or at least minimize, this pre-emptive inscription of the future, it is necessary to transform our understanding and engagement with hegemony itself. Rather than assume it as a battle between competing ideologies, it is key to view it as a process by which lack gives birth to abundance. In other words, where the absence of freedom in any given status quo catalyses the creation and experimentation with a wide range of new ones. It is in Derrida's words "to decide to change the terrain, in a discontinuous and irruptive fashion, by brutally placing oneself outside, and by affirming an absolute break and difference" (1982: 135).

Equally significant is the need for completely transforming our relationship with others. As mentioned above, Sartre has a quite ambiguous notion of what a free relation between people could and should look like. He is in turns dismissive, opportunistic, or overtly normative. While his notions of an "us" for itself is certainly interesting, it provides little blueprint or detail of what this actually entails. Derrida offers a clearer roadmap for what such a collective deconstructive engagement can be, in this regard, associated with the concept of "friendship" (Derrida 1997). He proposes the need for a deep "hospitality" towards the other, using the concept of friendship as a model. This "generosity" is not naïve nor is it equivalent to simple liberal values of tolerance. Instead it is radical commitment to understanding how the presence of others fills in one's own absences, creating the basis for a shared reinterpretation of reality. Hence, through our experience with others, we perfect our conception and experience freedom, a process which is never complete and forever open to the potentials held by new encounters. This echoes Sartre's own conception of a radical collectivity in the form of a "we":

> Yet it is nonetheless true that the "we" subject does not appear even conceivable unless it refers at least to the thought of a plurality of subjects which would simultaneously apprehend one another as subjectivities, that is, as transcendences-transcending and not as transcendences-transcended. If the word "we" is not simply a flatus vocis, it denotes a concept subsuming an infinite variety of possible experiences. (Sartre 1956: 413)

Freedom, therefore, is constantly present in its forever absence. It challenges us to deconstruct, reinterpret, and reconstruct our reality. It is an unfinished project that can never be contained by any mystical claims that its "true" or "objective" essence has been found. It is a "Freedom to be invented. Everyday. At least. And Democracy along with it" (Derrida 1992: 98). If it is intrinsic at all, it is in its haunting of society to explore its possibilities further. In doing so, we keep open the existential gap found in the eternal tension between what is and is not yet. The spectre of freedom troubles established social orderings and disturbs our bad faith.

References

Arthur, M., & Rousseau, D. (2001). *The Boundaryless Career*. New York, NY: Oxford University Press.
Barton, D. (2011, March). Capitalism for the Long Term. *Harvard Business Review*.

Beauvoir, S. (1948). *The Ethics of Ambiguity*. New York: Citadel Press.

Bell, L. (1989). *Sartre's Ethics of Authenticity*. Tuscaloosa, AL: University of Alabama Press.

Bingham, C. (2007). Derrida on Teaching: The Economy of Erasure. *Studies in Philosophy and Education, 27*(1), 15–31. https://doi.org/10.1007/s11217-007-9044-4.

Bloom, P. (2017). *The Ethics of Neoliberalism: The Business of Making Capitalism Moral*. London: Routledge.

Busch, T. (2002). Thomas W. Busch – Gadamer and Sartre on Self-Transformation. *Symposium, 6*(2), 195–202.

Choat, S. (2016). Marxism and Anarchism in an Age of Neoliberal Crisis. *Capital & Class, 40*(1), 95–109. https://doi.org/10.1177/0309816815627751.

Choudhary, K. (2007). Globalisation, Governance Reforms and Development: An Introduction. In K. Choudhary (Ed.), *Globalisation, Governance Reforms and Development in India*. London: Sage.

Derrida, J. (1974). *Of Grammatology*. Baltimore: John Hopkins University Press.

Derrida, J. (1978). *Writing and Difference*. London: Routledge.

Derrida, J. (1981). *Positions*. Chicago: University of Chicago Press.

Derrida, J. (1982). *Margins of Philosophy*. Chicago: University of Chicago Press.

Derrida, J. (1992). *The Other Heading: Reflections on Today's Europe*. Indiana: Indiana University Press.

Derrida, J. (1994). *Specters of Marx: The State of the Debt, the Work of Mourning, & the New International*. London: Routledge.

Derrida, J. (1997). *The Politics of Friendship*. London: Verso.

Derrida, J. (2002a). *Acts of Religion*. London: Routledge.

Derrida, J. (2002b). *Who's Afraid of Philosophy? The Right to Philosophy*. Stanford: Stanford University Press.

Derrida, J. (2004). Autoimmunity: Real and Symbolic Suicide. In G. Borradori (Ed.), *Philosophy in a Time of Terror Dialogues with Jürgen Habermas and Jacques Derrida*. Chicago: University of Chicago Press.

Foner, E. (1999). *The Story of American Freedom*. New York: W.W. Norton and Company.

Goggin, G. (2006). *Cell Phone Culture*. Hoboken: Taylor and Francis.

Gyllenhammer, P. (2015). Progress and the Practico-Inert. *Sartre Studies International, 21*(2), 2–32.

Hall, D. T. (2004). The Protean Career: A Quarter-Century Journey. *Journal of Vocational Behavior, 65*(1), 1–13.

Hall, P., & Soskice, D. (2004). *Varieties of Capitalism*. Oxford: Oxford University Press.

Leonard, M. (2014). Rage Against the Machine: The Rise of Anti-Politics Across Europe. *New Statesman.*

Monbiot, G. (2016). Neoliberalism – The Ideology at the Root of All Our Problems. *The Guardian.*

Obama, B. (2009). *New Conference for London Summit.* Delivered at Excel Center in London, UK.

Pasquino, P. (1991). Theatrum Politicum: The Genealogy of Capital-Police and the State of Prosperity. In C. Burchell, C. Gordon, & P. Miller (Eds.), *The Foucault Effect: Studies in Governmentality* (pp. 105–118). Chicago: Chicago University Press.

Reich, R. (1987, May). Entrepreneurship Reconsidered: The Team as Hero. *Harvard Business Review, 65,* 22–83.

Sanyal, K. (2014). *Rethinking Capitalist Development: Primitive Accumulation, Governmentality and Post-Colonial Capitalism.* London: Routledge.

Sartre, J. (1956). *Being and Nothingness.* New York: Gallimard.

Sartre, J. (1963). *Search for a Method.* New York: Vintage Books.

Sartre, J.-P. (1948). *Anti-semite and Jew.* New York: Schocken Press.

Shamir, R. (2008). The Age of Responsibilization: On Market-Embedded Morality. *Economy and Society, 37*(1), 1–19.

8

Reinvesting in Good Faith: The Radical Promise of Existential Freedom

At the beginning of the twenty-first century, humanity finds itself at a profound existential crossroads. Down one path lies our continued bad faith in the free market—a religious devotion to an economic dogma that not only spells our material ruin but robs us of our fundamental agency to reinterpret and transform our world. Down the other is the individual and collective freedom to shape our historical destiny and expand the horizons of what is socially possible. It is our choice and one that will define our shared future.

Present-day capitalism is seeking to fundamentally rob us of this freedom by marketization our future. In the prescient words of scholar Will Davies (2014: xii):

> But surveying our model of political economy in the wake of the global financial crisis, there is an eerie feeling that for all the talk of uncertainty today, things appear to have become a tedious and painful procession of the same. Speculation and imagination regarding the future have free reign, so long as they are not turned upon the political conditions that seek to guarantee and secure them within certain limits. While we view our own

© The Author(s) 2018
P. Bloom, *The Bad Faith in the Free Market*,
https://doi.org/10.1007/978-3-319-76502-0_8

fates as subject to unpredictable buffeting by competitive forces, the "game" within which these forces operate feels utterly permanent … We must be ready for anything: but somehow this is never cause for hope of real change.

It is not so much that there will be no tomorrow in this scenario, rather that anything that comes after the present will naturally reflect dominant free market values. We are therefore existentially fated to become little more than "me-as-objects" to a permanent nightmare of a neoliberalism that is determined to last forever whatever the human costs.

The dominant response to this historical predicament has been the unappealing options of embracing what can be termed a "genuine cynicism" or a "hopeful nihilism" (Bloom 2016). The former thus would genuinely like a different world but remains resigned to pragmatically accept the status quo as inalterable. The latter similarly wants radical change, yet cares for little what this may turn out to be and who it may ultimately harm. If the landmark 2017 presidential election is any predictor of our upcoming political prospects then a radical existential intervention rather than a radical existential reboot is clearly urgently needed.

What is crucial for such a global and transformative task is our willingness to give up our bad faith in the free market. This involves more than simply rejecting its claims of objectivity or orthodoxy. By contrast, it is a critical reflection of how and why we deceive ourselves in its name—continuing to believe fervently in its salvationary power when all signs point to the fact that it is taking us down the road to future capitalist tyranny. In the present age, it is not socialism that threatens our fundamental freedom to think, act, and live. Instead, it is our investment in a dogmatic market freedom whose destination is our continued existential bondage.

In its place we must reinvest in the good faith of our own foundational freedom to consciously interpret our reality and reshape our world and toonce more place the openness of existence before the certainty of our socially constructed essences. It is to deconstruct and revolutionize the "situation" of our present freedom and in doing so open up the possibility for new freedoms to be conceived and materially flourish. It is an opportunity to transcend market freedoms in the search for new meanings and forms of liberation.

Market Reactions

The free market breeds increasingly impassioned responses from people—ones that are both positive and negative. The mere fact though that is so firmly on the public radar around the world reveals just how open it is to our shared existential reconsideration. While these challenges have catalysed efforts to genuinely transform the social in quite progressive ways, it has also revealed the threat of the return of the market under the guise of radical freedom.

If nothing else, the last decade highlights the resurgent possibilities of politics generally. In the face of a seemingly immovable neoliberal consensus, ordinary people have risen up to protest for change. On the Left this is apparent in the vibrancy of the anti-globalization movement as well as popular revolts from the Arab Spring to Momentum in Britain. On the Right this is witnessed in the global nativism sweeping much of the world, a populist upsurge for the right of nations and "native people" to once more control their own borders, historical fates, and "way of life". To this end

> Europe is not so much witnessing the mainstreaming of radical right policy agendas and discourses or the shifting of the political spectrum to the right, as it is entering into a nativist period in which constructions of belonging and policies regarding immigration and cultural diversity are being shaped by a new cleavage between 'natives' and 'non-natives.' Nativism also helps broaden the analysis of Islamophobia and xenophobia by highlighting the new ways in which a self-appointed 'native' society defines its values vis-à-vis the fundamental threat posed by a particular immigrant group or ethnic community. In defining these values, nativism can, and is, increasingly taking on civic characteristics, such as a defence of secularism, gender equality, or the rights of sexual minorities. (Guia 2016: 1)

While both these politics seek to upend the current order, each can be countered on existential grounds—though to vastly different degrees. The Conservative rejection of corporate globalization is certainly a rebuking of the present Gods of the international free market. However, it is also a reaffirmation of previous essential identifications—a literal and figurative attempt to return to a fanciful nostalgic past (see Kenny 2017). It is trading of one bad faith for a more toxic and repressive other. The progressive rejoinder is much less existentially suspect in this respect. Nevertheless, it also runs the risk of reviving utopian like "social democratic" past with a similar religiosity now granted to the free market.

The existential task is to explore what new freedoms our present situation provides and recalibrate our interpretation of the world to reflect these emergent possibilities. It does not mean eschewing all that has come before. Conversely, it is to use existing freedoms as a jumping-off point for a still undetermined future rather than an anchor dragging us back to an already decided past. Hence, liberal values of free speech and human rights must not be abandoned, casualties of an understandable desire to make a complete break with an oppressive and tyrannical status quo. Yet they should also be forward-thinking rather than backward looking, how can they inform and be made part of the efforts to transcend being as it is in order to continue to explore what it may still become.

It is exactly at this point that the traditional critique of existentialism as normatively ambivalent must be firmly addressed. Put differently, it is commonly said that if existentialism provides a compelling ethos of freedom, its own moral standing is decidedly lacking. It can be critically questioned given that

> If there is no determinate human essence, then there can be no univocal conception of what it is, exactly, for human being to achieve itself. But then there can be no univocal standard to determine whether an action is good or evil. In other words, if there is nothing to advance whether an action advances or detracts from the self-achievement of the agent in question, then no action is any more good than it is evil. The question again is whether existentialism can escape this predicament. Or, does the existentialist's denial of human essence destroy the possibility of morality? (Tanzer 2008: 55)

Its emphasis on the fallibility of traditional meanings and freedom to discover new ones can, of course, lead humanity down potentially rather dark paths. If all is permissible, then nothing can be morally judged. Nevertheless, this interpretation misses a key element of existentialism—

especially as it has attempted to be presented throughout this work. Notably, we should evaluate freedoms not merely on their substantive consequences or intrinsic values but also to the extent to which they contribute to our fundamental existential freedoms. The genocides and oppressions of the past century—from the gulags of really existing Communism to the concentration camps of fascism to the ongoing mass incarcerations and material deprivations of the global capitalism—certainly reflected the human capacity to remake the world yet only as attached to their own chosen ideological Gods.

The threat of the market then runs so much deeper than its destructive daily injustices or mass inequities and impoverishment. It is the risk of essentialism, of sacrificing our deeper existential freedom on its pseudo-religious altar. The reactions it has given birth to not only reinforce its own values but also the underlying commitment to an inherent essentialism upon which it has materially and ideologically built our modern world upon. The latter half of the twentieth century witnessed an explicit commitment by global elites—in rhetoric if not necessarily in practice—to distance market freedoms from its bloody colonial, racist, and patriarchal beginnings. It was pointing to a brave new capitalist world where everyone can not only succeed but also shape their own destiny regardless of their background. The reality was a reinvestment in an oppressive market fundamentalism. One that increasingly drew its strength from these once thought moribund essentialized identities while rejecting any future out of hand in which humans could craft their own existence free from supposed objective laws or their perceived nature.

With the election of Trump and the advent of Brexit as well as the surge in popularity of far-right politicians internationally, the market has certainly reacted to the present "challenge of freedom". It has reinvigorated broad swathes of people to at the very least once more feel liberated to topple an existent order, wrapping its reified values and exploitive practices in the new clothes of "radical freedom". This is a quintessential bad faith, as it is deceiving ourselves to believe that we are in fact the agents of our actions, the choosers of our fate, all the while accepting an existence dedicated to the preservation of a divine or transcendental force seen as ultimately responsible for guiding and deciding our individual and collective destiny.

Having Good Faith

At stake then, is how can we escape our bad faith? More precisely, how can we as conscious beings reclaim and preserve our existential birthright of freedom? Key to addressing these questions is to better understand what constitutes good faith. It is crucial to unravelling the radical implications of nothingness, to be revolutionary in our intention and actions rather than pragmatic in our resignation that what is currently absent can never be made present.

Sartre, in this regard, does not merely pose the problem of "bad faith". He also holds out the potential for having "good faith". To this end, "Good faith seeks to flee 'not believing what one believes' by finding refuge in being" (Sartre 1956: 70). Tellingly, he at points seems to be rather indifferent to the emancipatory potential of good faith. However, while always fraught, it does point to the ability to have faith in being faithless to any essentialized version of Being and optimistic about our capacity as the makers of our own meaning in an otherwise indifferent universe. To believe fully in our freedom and act accordingly. Hence according to leading Sartrian interpreter Ronald E. Santoni (1995: 71), good faith is "an open critical attitude toward evidence" and it "accepts the anguish associated with its freedom … and assumes responsibility for the authorship of its beliefs and actions" (Ibid.: 80). Authenticity, meanwhile, is "a willed 'conversion' from one's 'natural' attitude of trying to flee one's freedom to a lucid recognition of one's situation, [and] also invokes the acceptance, affirmation, valorizing, and living of one's freedom, one's 'non-coincidence' of being, one's ontological abandonment to responsibility" (Ibid.: 111).

The ultimate aim, hence, is to make "being-for-itself" a condition of "being-in-itself". Put differently, to render our existential freedom as so pervasive that it simply becomes a part of any and all of our existences. Yet the failure for this to ever occur should not lead to desolation. Instead it is precisely in this "existential gap" that new freedoms emerge and fresh impulses for rediscovering the possibilities of existence are realized. In this respect, it is to "authentically" embrace our freedom and take responsibility for it.

Tellingly, most existential analysis begins and largely ends at the level of the individual. It is about personal freedom to choose how one lives and deals with a world void of any inherent meaning. However, as discussed in the previous chapter, Sartre certainly introduces a collective idea of existential freedom. This is a radical gesture towards the capacity of individuals to come together with the unified purpose to form their own foundations and practices for guiding their shared existence. Moreover, it reveals the limitations of thinking solely individually in terms of capturing our existential freedom. Notably, it is with others that the realization of our absences, our non-being, becomes more comprehensive and our capacities to escape our situation and concretely explore the possibilities of nothingness are enlarged. It is therefore the recognition of a continual desire to create not only a "being-for-itself" but a collective "society-for-itself".

These shared existential longings bring us too then the "responsibility" of such fundamental freedom. Traditionally, liberty is associated with a sense of obligation. What then is our associated duty in relation to existential freedom? It is first and foremost to reject any and all attempts at establishing permanent foundations. It is a commitment to the contingency of our reality and ours as well as others ability to reinterpret it. Yet it also exceeds conventional liberal tolerance, as it is a dedication to both locally and globally critically interrogate what is preventing such freedom from being realized and seeking intervention against and undermine such barriers. This means understanding the material and discursive "challenges to freedom" so that they can be concretely transcended.

Existential freedom then must become a guiding principle for founding and refounding our personal and shared existence. Crucial to such a project is to consistently transform the "facts" of Being into "facticities" of existence that can and must be overcome. The French philosopher Rancière (2008), to this end, posits that the ethos and demand for "equality" always unsettles any and all social orders, leading to the struggle against the "policing" of its entrenched power relations. In an analogous fashion, freedom "haunts" our existence, undermining its claims to permanency and seeking to move beyond its pretensions of essentialism. It is the constant and diverse means through which we overcome the "policing" of Being, emancipating it for the liberation of a radical nothingness composed of our own ability to make and remake the world in as yet unimaginable ways.

A Freedom to (Not) Believe in

However, if the potential of non-being is our central tenant, then are we left with nothing substantial upon which to base this good faith? Faith even in nothing requires paradoxically a belief in something. Further, if all is ultimately meaningless and fleeting, why invest in any social order at all? Are we not truly "condemned to freedom" as Sartre suggests, fating us to an existence of anguish over our own impotence in the face of an impersonal and uncaring university? It is, accordingly, imperative to note that this existential commitment to freedom is not merely one of naïve optimism. Rather, it is a direct engagement with our fundamental condition as conscious living beings.

As expressed throughout this book, the rejection of any essential human foundations does not entail that there is nothing fundamental about our existence as such. Put differently, Sartre repeatedly asserts that it is our consciousness that defines our experience of the world as "free beings" in so much that it introduces the very notion that there is a lack and the potential of "being-for-itself". In this sense, by recognizing ourselves as being conscious, we are already latently believing that our present being is not exhaustive. There is thus an always positive component to this continual negation, the assertion that we are more than our present situation, and the recognition that through consciousness we can reinterpret our existence differently.

Yet there is a strong belief component to this fundamental freedom. Notably, we must have faith that something more is possible and that we can be the agents of this change. Again, this good faith is not a matter of simple idealism or romanticization. Rather, it reflects the productive tension between consciousness and knowledge. The former is an awareness of "what is" and as such its corollary of "what is not". The latter is what Sartre refers to as an "intuition" of our freedom and that what is currently nothing can one day be something. More precisely, it is the "consciousness of a thing" and as such an engagement with "non-being". He writes hence "we should note furthermore that this non-being is implied a priori in every theory of knowledge" (Sartre 1956: 173). Accordingly, the negation of consciousness and the assertion of knowledge can combine to give us good faith in our continual ability to transcend the present being.

In practice, this means be willing to affirmatively say no. Using different language, to critically reject what currently is for the prospect for what may still emerge. Significantly, the existential "no" to a given "situation" necessarily also implies a yes. When forsaking one course of action or an entrenched identity, there is always a positive reason motivating this negative event. It is a consciousness that this essence is lacking and the knowledge that existence can offer us something different and better. Further, these are always embedded recognitions—thus the rejection of, say, racism or classicism is inspired by the awareness that equality is lacking and the intuition that a more equal society can be created, a prospective world that is known in principles but not necessarily details. Therefore, in saying no there is the affirmation that a different as yet undiscovered existence is both possible and desirable.

Existential freedom then is the freedom not to believe in what is and to believe in what is not yet. It is a suspicion of essence and a celebration of existence. It is the good faith that nothing is permanent. No social reality is inalterable and what is not present will always haunt what is present. Freedom is eternally incomplete, inexhaustible, and positively non-existent. Existentialism, thus, offers us perpetually a radical freedom to not believe in.

Reactivating Freedom

Existential freedom though is not just about belief—it is about acting on these beliefs. We are indeed "condemned to freedom". Yet what this means is much more complex than perhaps gleaned at first glance. It refers neither—or not wholly—to a fatalistic acceptance that we must choose in a universe that provides us with very little choice nor to a full-scale optimism that we can transform the world at all times according to our own desires. By contrast, it points to rich ways freedom is fundamental to both our personal existence and existence generally. In this respect, freedom continually acts upon us and must be constantly reactivated by us.

A crucial but too often overlooked insight is that freedom is always present. The very fact of consciousness is the recognition of this persistent freedom—the realization that things could be otherwise imply therefore that one has a choice. At the very least it is a selection of interpretation (which aspects of a situation one prioritizes). Comprehensively, it is also a bounded freedom of action, you could always choose otherwise regardless of how dire one's options. Even in the face of impending execution, one can choose the manner of their death—whether they face it stoically, impassionedly, and so on.

However, this book marks a stronger claim about the permanence of freedom. Being also confronts us with an "existential gap", a chasm between our desire to shape our environment and our inability to fully do so. This is unavoidable, though it may be felt more strongly at certain times and over particular events than others. Further, every expression of existence produces its own unique existential gaps and commonly its own essentialized freedoms for seeking to overcome them. Hence, freedom is eternally present, at once reflective of and shadowed by its fundamental existential origins.

Capitalism, therefore, creates market derived "gaps" of freedom—linked to inequality, our desire for happiness and connection, impoverishment—of which market freedom is put forward as the answer to. This simultaneous problem and answer to the situational desire for freedom are mutually reinforcing. More precisely, the dominant philosophy at the time will pose the "problems of freedom" that its own essentialized freedom is supposedly best able to address. Hence, under capitalism, the need for greater efficiency, innovation, and productivity are constantly promoted as the key to enhancing our freedoms—all of which fit close to perfectly with established conceptions of market freedom.

This reveals an underlying tension within Sartre's account of existentialism. Notably, while existence precedes essence, we most often consciously experience it in the reverse order. Our first experience of being conscious is in learning "who we are", reinforcing an inherent sense of self from which we can either embrace or seek to escape. In this respect, for there to be a negation (a "what is not") there must first be a concurrent substantive understanding of being (a "what is"). Yet the lack of consciousness, the persistent awareness of non-being, is constantly shaped by and partially filled by prevailing forms of freedom. Existentialism freedom then is necessarily a double move—at once emancipating ourselves from an existent freedom so that we can be liberated to conceive and enact alternative ones.

Theoretically, this speaks to growing distinction between politicization and depoliticization. The latter refers to the naturalization and subtraction of any politics from key social institutions and decisions such as the economy or policing (see Ferguson 1994). Politicization, on the other hand, denotes the reactivation of this social contingency—the ability to contest and change existing socio-political relations (Glynos and Howarth 2007). This represents a more thoroughgoing separation of the political from domination. Post-foundational thinkers, according to Marchart (2008), share a belief in an "ontological difference" where emerging "political" forces continually threatens institutionalized formal "politics". Freedom is manifested in analogous ways, as an essentialized freedom will be challenged by an existential freedom.

Significantly, freedom must therefore be persistently reactivated existentially. Critical to doing so is radicalizing an "existential gap" so that it challenges rather than reproduces an existent status quo and its supportive account of objectified freedom. Our freedom is, therefore, constantly being interrupted by itself, unsettled by reinterpretations of what it is not as well as what it could and should be. Thus to authentically "be free" it is crucial to set Being free from its perceived essences.

Reinvesting in Our Good Faith

Central to capitalist existence is the need for investment. We are asked to invest our money, time, and talents in its success. The exchange for such investment is the ability to freely choose how we live and pursue happiness according to our personal desires. However, there is an increasing awareness that this investment is rapidly losing its value. It is an investment in our bad faith—the continual clinging to a sense of freedom that we know to be ultimately repressive rather than liberating.

Crucially, freedom demands to be invested in generally. We effectively attach our sense of self, our aspirations, and our desire for agency to existing commonly essentialized forms of freedom. Alternatively, at our most existentially radical moments, we are tasked with investing our critical imagination and capabilities in a freedom yet to come. This entails devoting our mind, time, energy, and abilities into negating the present and founding new ways for living. It means, therefore, taking seriously our conscious awareness of that something is lacking and having faith in our knowledgeable intuition that there is still something more to our existence.

For most of the present period, we invested ourselves almost completely in market freedoms. Ostensibly, this is linked to the belief that the free market is our best and only chance for personal fulfilment and collective prosperity. It involves accepting "market rationality" as the end all and be all for transcending our present condition, for going beyond our existing "situation". We invest in market solutions—whether related to employability or consumption—to overcome our experience of lack. And just as significantly, it all assumed that no matter what you dreamed of achieving—whether materially, ethically, or both—that doing so meant embracing our human "nature" as a market subject.

Yet this investment in market freedom has brought progressively diminishing returns. Rather than inspiring confidence in our abilities and empowering us to choose the life we would want, it largely endows us with ever greater anxiety and disappointment about our lack of agency to change either ourselves or our society. It also puts forward seemingly unacceptable facts about our current human condition—that inequality is inevitable, that we can no longer afford high-quality universal healthcare, or the best education for all, and so on—that reveal not only its fundamentally lacking qualities. New freedoms are clearly needed and urgently so.

Required, in turn, is the creation of a radically different political economy. As discussed earlier in this book, Foucault distinguishes between the economic which expands our capabilities and politics which directs them in certain proscribed directions. In the wake of the financial crisis and the lack of genuine recovery for the majority of the world's population, it is easy to understand why alternative economic models would therefore be so suddenly appealing to a growing many. Existentially, this reflects the significance of continually preparing for non-being, developing new skills and understandings so that it becomes possible to achieve the previously considered impossible.

In more concrete terms, this reflects how imperative it is to fundamentally broaden our economic horizons, to literally free our minds of market freedoms to imagine different and more empowering social possibilities. It is telling that there is a growing fear over the coming "rise of the machines"—the use of "smart technology" to materially displace us and further socially alienate us. This soon to be here "industry 4.0" is met most often with a sense of dystopian dread. As a recent *Financial Times* article proclaimed "Fears of automation overshadow our road to the future. Our anxiety about technological innovation has changed little in the past 120 years" (Jezard 2017: N.P.).

What makes this so revealing is not that such fears are not to a degree warranted but rather that they are warranted precisely because of how pernicious and limited we have become due to our overwhelming investment in the free market. All new advancements are met with suspicions, as potential sources for our further exploitation and repression. Instead, it is absolutely vital to embrace the liberating potentials of this future, choosing a radically different path and developing the social capabilities for its realization. In the words of noted critical scholar Erik Olson Wright (2016: 1) in his increasingly influential text "How to be an Anti-capitalist"

> We live in a world where capitalism, as a system of class relations and economic dynamics, creates enormous harms in the lives of people ... While there is widespread recognition of these problems, nevertheless the idea of a viable alternative to capitalism that would avoid these harms and make life genuinely better seems quite far-fetched to most people. In part the issue is simply skepticism that an alternative—even if it can be imagined—would actually work in practice. But even among people who do believe in the viability and desirability of a democratic, egalitarian, solidaristic alternative to capitalism, there is little confidence that an emancipatory alternative to capitalism is politically achievable. The problem here is not mainly the ability to imagine the goal of an emancipatory social transformation; the problem is imagining a strategy to realize that goal—how to get from here to there.

If we want to make our tomorrow worthwhile it is, therefore, key that we reinvest in our own good faith. It is the jettisoning of our market essence for a new transformative existence. It is to experiment with how we live and how we seek to overcome our always present existential gaps. The challenge of modern freedom is one of total reimagination, to become optimistic heretics against an orthodox order suppressing our potential.

The Radical Promise of Existential Freedom

The hope of a new millennium is to have the freedom to universally reconfigure our concrete existences. Specifically, it is to never foreclose the potential of Being to become different versions of "beings" for different beings.

This revolutionary good faith allows us to see beyond essentialized freedoms in order to conceive and live out alternatives. Existential freedom identifies a range of absences so that we make better understand what is absent and make it a concrete reality. It is to, in short, highlight what Being is so that we may explore its multiple and eternal forms of nothingness.

Derrida refers, in this regard, to the "promise of democracy". Specifically, it reflects a democracy "to come"—one always on the horizon pushing us forward to perfect our eternally imperfect times. It must be the "opening of this gap between an infinite promise [...] and the determined, necessary, but also necessarily inadequate forms of what has to be measured against this promise" (Derrida 2008: 65). Hence, the "very motif of democracy", its exact "possibility", is found in "the duty of democracy itself to de-limit itself" (Derrida 2005: 105). This radical conception of democracy thus simultaneously affirms democracy as an ideal while intervening against any and all existing examples of it. If this sounds counterproductive, it is not meant to be. Rather it is a celebration of democracy's inherent openness and a demand that it never stop exploring possibilities for manifesting such values. In doing so, it maintains our faith in the possibilities for continually deconstructing and reconstructing our world. Consequently for Derrida (1994: 76)

> what remains irreducible to any deconstruction, what remains as undeconstructible as the possibility itself of deconstruction is [...] perhaps even the formality of a structural messianism, a messianism without religion, even a messianic without messianism, [...] an idea of democracy which we distinguish from its current concept and from its determined predicates today.

What is promised by existentialism is exactly nothing. More precisely, it offers out the hope that there is still more to come, that being is always subject to change, that non-being represents not our total annihilation but an exciting potential beyond our current imaginations. It is the promise of being eternally able to give an affirmative no to our present situation. The intuition that what appears to be nothing in fact contains multitudes to be discovered. And a recognition that none of the decisions are ever final, that we can always choose our lives anew whatever our past. Quoting renowned Democratic theorist Aletta Norval (2004: 147)

> If, however, one of the characteristics of the terrain of the undecidable is that it resists closure, as we have seen, it is also that which inaugurates the

need for a certain 'decision', for it marks an irreducibly plural terrain, a terrain in which identity is still at stake waiting to be inscribed.

Existential freedom is found in actively doing absolutely nothing. Again to be clear, this is not a resignation to passivity. Instead, it is a decision to be in ways that have not been before. To explore, tentatively with great passion, persistently and at points rapidly, that which does not yet exist. "What must be thought is this inconceivable and unknowable thing", quoting Derrida (2006: 152), "a freedom that would no longer be the power of a subject, a freedom without autonomy, a heteronomy without servitude, in short, something like a passive decision".

Foucault, for his part, mentions the need for "thinking otherwise" and by association the radicality of "doing otherwise". He proclaims "There are moments in life where the question of knowing whether one might think otherwise than one thinks and perceive otherwise than one sees is indispensable if one is to continue to observe or reflect." This is sage advice and one that despite being necessarily local in its scope can have quite profound and radical general implications. What existentialism adds to this call is to embrace the freedom of nothingness, to develop what is initially otherwise into a journey into the heart of non-being, exploring that rich absence in our given presence. While involving reflection, this journey is not merely abstract or meditative. Rather it is a practical ethics to choose to constantly "reproduce" our daily existence instead of simply reproducing any essential being or entrenched social order.

Required is a radical commitment to the vast potential held in our nothingness. We must negate our present to avoid our own future existential annihilation. This radical commitment to existential freedom is especially significant in the face of a religious revival worshipping the free market and its capitalist Gods. Amidst all the present upheavals and fears by elites that "our very way of life" is suddenly under attack, a fundamental truth is risked being missed. It is that we are the ultimate creators of our world, that our reality is ours to make and remake, that there is no "invisible hand" directing our lives or objective laws determining our historic fate. And most important that Being is never finished becoming and freedom never stops. The greatest opportunity we still presently have for salvation is to give up our bad faith in the free market in order to realize the radical promise of existential freedom.

References

Bloom, P. (2016, November 9). Trump and the Triumph of Hopeful Nihilism. *The Conversation*.

Davies, W. (2014). *The Limits of Neoliberalism: Authority, Sovereignty, and the Logic of Competition*. London: Sage.

Derrida, J. (1994). Spectres of Marx. *New Left Review, 205*, 31–58.

Derrida, J. (2005). *Rogues*. Stanford, CA: Stanford University Press.

Derrida, J. (2006). *Politics of Friendship*. London: Verso.

Derrida, J. (2008). *Ghostly Demarcations*. London: Verso.

Ferguson, J. (1994). *The Anti-Politics Machine: Development, Depoliticization, and Bureaucratic Power in Lesotho*. Minneapolis: University of Minnesota Press.

Glynos, J., & Howarth, D. (2007). *Logics of Critical Explanation in Social and Political Theory*. New York: Routledge.

Guia, A. (2016). *The Concept of Nativism and Anti-Immigrant Sentiments in Europe*. EUI Working Papers, 20.

Kenny, M. (2017). Back to the Populist Future? Understanding Nostalgia in Contemporary Ideological Discourse. *Journal of Political Ideologies, 22*(3), 256–273. https://doi.org/10.1080/13569317.2017.1346773.

Jezard, A. (2017). Fears of Automation Overshadow Our Road to the Future. *The Financial Times*, 3 May.

Marchart, O. (2008). *Post-Foundational Political Thought*. Edinburgh: Edinburgh University Press.

Norval, A. (2004). Hegemony After Deconstruction: The Consequences of Undecidability. *Journal of Political Ideologies, 9*(2), 139–157. https://doi.org/10.1080/1356931041000169187.

Rancière, J. (2008). *Disagreement*. Minneapolis [u.a.]: University of Minnesota Press.

Santoni, R. (1995). *Bad Faith, Good Faith, and Authenticity in Sartre's Early Philosophy*. Philadelphia: Temple University Press.

Sartre, J. P. (1956). *Being and Nothingness*, Trans. Hazel E. Barnes. New York: Philosophical Library.

Tanzer, M. (2008). *On Existentialism*. Belmont, CA: Thomson Wadsworth.

Wright, E. (2016). How to Be Anti-Capitalist. *Jacobin*.

Index

NUMBERS AND SYMBOLS

2008 Financial crisis, vi, 1, 13, 21, 41, 43, 165–166

A

Abandonment, 29, 35, 176
Absence, 75, 78, 96–98, 103, 104, 110, 159, 160, 164–168, 177, 184, 185
Affect, 8, 44, 94, 95, 99, 111, 112, 128
Althusser, Louis, 32, 54, 134
Anguish, 28, 29, 35, 46, 98, 105, 176, 178
Annihilation, 74, 97, 98, 109, 111, 184, 185
Anxiety, 2, 20, 36, 45, 76, 93–95, 110, 129, 165, 182
Austerity, vi, 35, 42, 44, 66, 69, 95, 151

B

Being, 1, 19, 41, 65, 91, 117, 147, 173
Being and Nothingness, 14, 15, 27, 77, 86, 91–113, 131, 155
"Being-for-itself", 133, 176–178
"Being-in-itself", 28, 77, 95, 125, 133
Bell, Linda, 153

C

Capitalism, vi, 1, 3, 6–9, 12, 14, 15, 19–23, 32, 35, 41–63, 68, 69, 82, 84, 86, 91–95, 97, 106–112, 121, 123, 134, 146, 147, 166, 171, 175, 180, 183
Catalano, Joseph, 96, 97
Class struggle, 16, 50, 56, 59
Collective Life Project, 151–156
Communism, 23, 24, 50, 67, 136, 167, 175

© The Author(s) 2018
P. Bloom, *The Bad Faith in the Free Market*,
https://doi.org/10.1007/978-3-319-76502-0

Competition, 55, 63, 67, 80, 81,
  92, 164
Corbyn, Jeremy, 61
Critchley, Simon, 45
Critical theory, 13, 107

D

Davies, William, 119, 171
De Beauvoir, Simone, 155
Deconstruction, 12, 13, 112,
  145–168, 172, 184
Democracy to come, 161
Derrida, Jacques, 15, 158–165, 167,
  168, 184, 185
Desire, v, vi, 3, 10, 13, 20, 30, 33,
  34, 36, 53, 55, 59–61, 66, 67,
  70, 76, 80, 82, 83, 85, 86, 91,
  94, 95, 99, 100, 102–104,
  106, 108, 111, 112, 118, 122,
  123, 127–129, 133, 138, 140,
  145, 148, 150, 152, 153, 174,
  177, 179–181
Despair, 28, 97, 153
Dialectic, 43, 48, 52–57, 63, 154
Difference, 22, 30, 38, 54, 59, 82,
  96, 125, 133, 137, 147,
  158–160, 163, 167, 181
Discipline, 37, 66, 70, 80, 128, 132
Discourse, 5, 8, 10, 13, 20, 31, 33,
  45, 49, 59, 65, 66, 75, 81–83,
  92, 102, 103, 111, 126–130,
  133–136, 138, 139, 165, 173

E

"End of History", 1, 20, 42, 92
Enjoyment, 102, 104, 107, 108,
  110–113

Essence, 12, 13, 27, 29, 38, 53,
  55–57, 67, 79, 99, 149, 156,
  168, 172, 174, 179, 181, 183
Ethics (Protestant), 7, 70
Ethics of the drive, 111
Existence, vi, 1, 2, 6, 9–14, 24,
  26–32, 34–38, 48–50, 52–60,
  62, 68, 69, 71, 72, 74, 76–79,
  81, 83–86, 91, 95–100, 102,
  105, 106, 109–113, 117, 119,
  122–125, 130–134, 139–141,
  147, 149, 151, 153, 156,
  158–160, 164, 166, 172,
  175–181, 183, 185
Existential crisis, vi, 13, 41–63, 83
Existential freedom, vi, 1, 25, 31–33,
  35, 37, 41, 65, 91, 117,
  145–168, 171, 172, 174–177,
  179–185
Existential gap, 13, 19, 50, 52, 59,
  78, 132, 160, 168, 176, 180,
  181, 183
Existentialism, 9, 10, 12, 13, 15, 16,
  26–30, 46–49, 53, 76, 97,
  100, 102, 123–125, 130, 135,
  156, 158, 160, 174, 179, 180,
  184, 185
Existentialism and Humanism,
  26–29

F

Facticities, 14, 65–86, 113, 125, 177
Facts, v, 9, 11, 13, 14, 19, 42, 57,
  63, 65–75, 79, 82, 84, 86, 97,
  99, 103, 107, 113, 123, 126,
  132, 138, 152, 155–158, 160,
  161, 172, 173, 175, 177, 179,
  182, 184

Fantasy, 13, 15, 49, 67, 83, 84, 93, 95, 98–103, 105, 106, 110–112, 118, 122, 128
Financial economy, 7, 41
Financialization, 20, 43, 68
Foner, Eric, 4, 149
Foucault, Michel, 15, 16, 32, 119, 120, 126–133, 135–140, 185
Freedom, vi, 1, 19, 41, 65, 91, 117
Free market, v, vi, 1–16, 19–38, 41–44, 46, 49, 52, 54, 55, 57, 58, 60, 62, 63, 65, 66, 68, 69, 71, 72, 80, 82–84, 86, 92, 94, 95, 97, 106, 107, 109, 110, 117–141, 145–168, 171–173, 182, 183, 185
Friendship, 168

G

Glynos, Jason, 85, 111, 181
Good faith, 13, 16, 171–185

H

Hegemony, 9, 14, 66, 67, 74–76, 83, 86, 151, 165, 167
Heidegger, Martin, 30, 45, 46, 72, 79
Howarth, David, 75, 85, 181

I

Idealization, 23, 49, 56, 57, 60, 133
Ideology, v, 2, 8, 13, 21, 23, 32, 33, 47, 48, 54, 58, 68, 83, 111, 147, 167
Interpellation, 134, 135

J

Jameson, Frederick, 1, 84

K

Knowledge, v, 32, 42, 47, 48, 59, 61, 81, 95, 98, 119, 120, 123, 126, 129–132, 134, 136–140, 149, 151, 178, 179

L

Lacan, Jacques, 14, 16, 100–102, 104, 108–111, 128
Laclau, Ernesto, 14, 75, 82, 83, 85
Law, the, 71, 110, 160, 165
Life Project, 15, 151–156
Logics, 22, 58, 67, 70, 72, 82, 85, 127, 136, 147, 159, 164

M

MacGilvray, Eric, 4
Market freedom, 1, 2, 4, 6, 8, 10, 12, 13, 19, 24, 26, 31, 32, 34–36, 46, 59, 63, 66, 67, 70, 74, 76, 79, 81, 82, 86, 95, 103–107, 118, 119, 127, 129, 146, 148–151, 164–166, 172, 175, 180, 182
Market fundamentalism, 6, 7, 43, 175
Marx, Karl, 21, 23, 44, 48, 50–52, 54–56, 108, 146, 167
Marxism, 13, 14, 16, 23, 27, 43, 45–52, 60, 62, 63, 123, 167
May, Theresa, 32, 37, 140, 146
Misinterpellation, 135
Momentum, 20, 173
Mouffe, Chantel, 14, 75, 82

N

Negation, 55, 56, 75, 96–98, 163, 178, 180
Neoliberalism, 1, 2, 11, 14, 20, 24, 25, 31, 32, 35, 42, 58, 65, 93, 117, 118, 122, 134, 139, 140, 147, 150, 151, 172
Non-being, 10, 77, 78, 96–98, 102, 104, 106, 107, 109–111, 113, 159, 160, 177, 178, 180, 182, 184, 185
Norval, Aletta, 184

O

Obama, Barack, 94, 166
Open Marxists, 55

P

Perfectibility, 147, 148, 161, 162, 164–166
Philosophy, 2, 5, 6, 11, 12, 14, 15, 22, 23, 26–28, 35, 37, 43, 47–49, 51, 55, 56, 62, 77, 97, 112, 120, 123, 124, 131, 137, 147, 156, 158, 159, 161–163, 180
Play, 27, 54, 82, 86, 104, 112, 125, 129, 148, 156–160, 166
Populism, 25, 41, 122
Possibility, vi, 2, 6, 9, 10, 12–16, 21, 23, 25, 27, 28, 30, 33, 43, 50, 51, 55–58, 60–63, 65–67, 70, 74–79, 81, 83–86, 91, 93, 96–98, 102, 104, 108, 111, 112, 120, 125, 130, 135–139, 141, 145, 148, 151–153, 155, 158, 159, 161, 164, 167, 168, 172–174, 176, 177, 182, 184
Post-foundationalism, 32

Post-structuralism, 2, 11, 36, 37
Power, 4, 22–25, 30, 32, 34, 37, 41, 51, 53, 55, 56, 59, 85, 101, 103, 105, 110, 119–124, 126–140, 147, 150, 151, 154, 172, 177, 185
Presence, 9, 73, 78, 96, 97, 100, 102, 106, 131, 150, 152, 156, 158, 160, 166, 168, 185
Production, 14, 50–53, 55, 56, 58, 59, 62, 78, 105, 121, 126–128, 132, 136
Promise, 1, 9, 13, 15, 21, 24, 30, 32, 36, 46, 55, 58, 65, 94, 107, 108, 121, 122, 148, 152, 162, 167, 171–173, 175–185
Psychoanalysis, 16, 101, 105, 111, 112

R

Real, the, 28, 78, 131, 157
Religion, 1, 7, 22, 43, 100, 184

S

Sanders, Bernie, 20, 61, 66, 91
Sartre, Jean Paul, v, vi, 12–15, 27–30, 38, 47–49, 51, 56, 62, 68, 72, 73, 77, 78, 93, 95–103, 105, 112, 124, 125, 131, 151–161, 163, 167, 168, 176–178, 180
Search for a Method, The, 47
Seriousness, vi, 156, 157
Situation, 42, 73, 85, 98, 131–133, 140, 151, 153, 157–159, 161, 172, 174, 176–179, 182, 184
Spectre, 15, 23, 94, 145–168

Stavrakakis, Yannis, 101, 104
Stiglitz, Joseph, 7
Subject, the, 8, 77, 101, 103, 107,
    108, 112, 117, 163
Subjection, 15, 122, 126–129,
    132, 141
Subjectivation, 91, 127–129, 131

T
Traversal, 111, 126, 163
Trump, Donald, 8, 91, 175

Z
Zizek, Slavoj, 102–104, 111

Printed by Printforce, the Netherlands